giving
nature
a home

Robins

Marianne Taylor

B L O O M S B U R Y
LONDON · NEW DELHI · NEW YORK · SYDNEY

giving
nature
a home

The RSPB is the country's largest nature conservation charity,
inspiring everyone to give nature a home so that birds and wildlife can thrive again.

By buying this book you are helping to fund The RSPB's conservation work.

If you would like to know more about The RSPB, visit the website at www.rspb.org.uk
or write to: The RSPB, The Lodge, Sandy, Bedfordshire, SG19 2DL; 01767 680551.

Bloomsbury Natural History
An imprint of Bloomsbury Publishing Plc

50 Bedford Square
London
WC1B 3DP
UK

1385 Broadway
New York
NY 10018
USA

www.bloomsbury.com

BLOOMSBURY and the Diana logo are trademarks of Bloomsbury Publishing Plc

First published 2015

British Library Cataloging-in-Publication Data
A catalogue record for this book is available from the British Library.

Library of Congress Cataloguing-in-Publication data has been applied for.

ISBN: PB: 978-1-4729-1211-4
ePDF: 978-1-4729-2266-3
ePub: 978-1-4729-1212-1

2 4 6 8 10 9 7 5 3 1

Design by Rod Teasdale
Printed in China by C&C Offset Printing Co., Ltd.

MIX
Paper from
responsible sources
FSC® C008047

To find out more about our authors and books visit www.bloomsbury.com.
Here you will find extracts, author interviews, details of forthcoming
events and the option to sign up for our newsletters.

Contents

Meet the Robin

If you are somewhere in the British Isles as you read this, the chances are that there is a Robin not far away from you right now. If it is daylight (and even if it is not, and the street lighting is on instead), you might well be able to hear one singing. Should you be near a garden you could well spot another busily foraging at the edge of the lawn. This unmistakable little character is probably our most familiar bird species, as well as one of our most beloved.

We know the Robin best as a garden and parkland bird, and it certainly thrives in those places. However, before people came along to create parks and gardens it was a bird of forest, woodland, scrubby ground and other well-vegetated habitats, and it is still common in such habitats today. It has one of the widest distributions of any British bird, occurring throughout Britain and Ireland, including most offshore island groups. It is most numerous south of Yorkshire and Lancashire, but its range extends to the far north of Scotland, and it is only missing from the highest and barest mountains and uplands. There are nearly eight million breeding pairs across the British Isles (6.7 million of them in the UK), making the Robin our second most numerous breeding bird species (the Wren holds the top spot).

On mainland Europe the Robin (or European Robin, to give it its full name) is a more secretive bird in most areas than it is in the British Isles. Only in northern France does it have a similar prominence in folklore and legend. Most European countries are also less densely Robin-populated, with the huge landmasses of Poland, Spain

Above: Its habit of perching on raised objects to survey the scene is an endearing Robin trait.

Opposite: No British garden is complete without its resident Robin.

and Finland, for example, holding about 1.5 million, 3 million and 3.3 million pairs respectively. However, the Robin does occur across nearly the whole of Europe, parts of North Africa, and a sizeable chunk of western Russia. BirdLife International estimates its total world population across this extensive range to be somewhere between 137 million and 332 million individuals.

In less enlightened times settlers from the British Isles often brought wild birds from their homeland to new lands, hoping that their presence would make things seem a bit more homely. Attempts were made to introduce Robins to Australia and North America, but no viable populations were established. This is in contrast to the House Sparrow and Starling, which are both now extremely numerous in both areas, to the detriment of native birds.

Below: A French Robin. In much of Europe, the closely related Black Redstart is a much more familiar garden bird.

National treasure

Unlike most nations, the United Kingdom has no official, government-endorsed national bird. However, when *The Times* newspaper polled its readers in 1960 to select an unofficial one, three front-runners soon emerged – the Red Grouse, Wren and Robin.

The Wren is probably Britain's most widespread bird, and perhaps deserved the title for that alone; it is also a familiar and much-loved garden bird with tremendous charm and energy wrapped up in a pint-sized package. The Red Grouse had perhaps an even more legitimate claim to the title at the time, not because of the Glorious Twelfth, the start of the shooting season for Red Grouse, but because it was the only species in the world to occur nowhere else (the vagaries of taxonomy have since revised its status to a mere subspecies of the continental Willow Grouse).

The Robin is found across a wide swathe of the European and western Asian mainland beyond the British Isles, but it has a particularly close connection with British cultural tradition, and its confident nature, sweet song, pretty plumage and perky demeanour have thoroughly endeared it to successive generations of Britons. The Robin won the vote by a landslide, beating the Red Grouse and Wren into second and third place respectively.

Above: Although Robins have a wide Eurasian distribution, only in the British Isles are they ubiquitous garden birds.

Above: The Red Grouse, like the Robin, is another British favourite bird, though for rather different reasons.

Vital statistics

Above: Robins and Great Tits are about the same size and weight, but in plumage and character could hardly be more different.

A typical Robin measures 14cm (5½in) from bill-tip to tail-tip at full stretch, and has a wingspan of 21cm (8¼in). It weighs between 15.5 and 23.5g (½–¾oz), averaging 18g. Females tend to be slightly shorter winged and shorter tailed than males, but are slightly heavier. Many other 'little birds' that share the Robin's habitat have closely similar average measurements – the House Sparrow is of the same length but rather heavier, the Dunnock has shorter wings but weighs slightly more, and the Great Tit's average measurements are almost identical.

The characteristic red breast (or to be strictly accurate red breast, face and forehead) is the adult Robin's most striking feature. Its actual tone is more rich orange than red, and when you look at a Robin front-on you will notice that the red area bulges on either side of the breast, like a heart shape that has been inverted. The orange also becomes paler just behind the eyes. The upperparts are a warm mid-brown. Often a vague bluish-grey line along the bird's side separates the red and brown areas. The underside below the breast-patch is dusky-whitish. The bill is blackish-grey, the legs dusky pink and the large eyes very dark brown (looking black at any distance).

The Robin has a smoothly rounded head and body shape, and proportionately rather long legs and shortish wings. It tends to have an upright stance, often with the wing-tips hanging lower than the tail. The tail itself often has a curiously 'stuck-on' appearance, especially when the body plumage is puffed up, making an already rather round-contoured bird look practically spherical. These features together convey a very distinctive outline, so that it is often easy to recognise a Robin even when its breast-patch cannot be seen (for example when it is viewed from behind, or even in silhouette).

Non-British subspecies

There are several subspecies of Robin across Europe and Asia besides the one that occurs in the British Isles (*Erithacus rubecula melophilus*). More than 10 others may exist, depending on which taxomonist you listen to, but they show only minor variations in size and plumage tones. The more northern and eastern forms tend to be larger and paler than the southern forms, and have relatively longer wings (an adaptation to their more migratory lifestyles). The most distinctive Robins come from the Canary Islands of Gran Canaria and Tenerife. They are strikingly different from the mainland birds, having white 'spectacle' markings around their eyes, pure white bellies, and smaller and brighter breast-patches. The two islands have slightly different forms, but collectively these are sometimes split as a separate species, known as the Canary Islands Robin (*Erithacus superbus*).

Above: The Canary Islands Robin, with its brighter and bolder plumage, deserves its subspecific name 'superbus'.

Male or female, old or young?

Many people who get to know their garden Robin are curious about whether it is a male or a female. Unfortunately, this is one detail that the otherwise very open-natured Robin will not readily disclose. Males tend to have larger and brighter breast-patches than females, more prominent greyish side-stripes, and longer wings and tails, but all these differences are small, with much overlap. Behaviour is not much help in identification, as both male and female Robins sing and fight for their territories. When a pair is together during the breeding season, the interactions of the birds should (eventually) reveal which is which – the female begs for food from her mate, a behaviour that is rather easier to observe than the rather hurried actual matings.

Above: The breast-patch is, on average, slightly more extensive in male Robins, but the difference is variable and subtle.

Right: Mottled plumage and no red breast helps keep fledgling Robins safe – from both predators and other Robins.

Below: Though as spotty as a young Robin, a juvenile Dunnock is greyer, with less prominent eyes.

Juvenile Robins, in their first set of feathers, are completely different from the adults, being warm brown all over with a liberal sprinkling of yellowish spots. At this age they can be confused with similarly spotty juvenile Dunnocks, but the latter are slightly more greyish in tone, have smaller and lighter coloured eyes, and differ in stance and motion. Frequent tail-bobbing is a strong clue that you have a Dunnock rather than a Robin.

Young Robins begin to develop adult plumage when they are about eight to 10 weeks old, and after another couple of months have entirely adult-like plumage. These young birds can still be distinguished from older birds up until their next moult the following year (if you look closely). They show a narrow yellowish wing-bar (formed by pale tips to the greater covert feathers), and their tail feathers have pointed rather than rounded tips.

Colours and curiosities

Above: No carotenoids in their diet results in white, rather than pink, flamingos.

Below: A Kingfisher's iridescent blue is a structural colour, rather than pigment-derived.

The colours in bird plumage are formed in two ways. Most matt colours are the result of pigments laid down in the developing feathers. In some cases these pigments are derived directly from a specific part of the diet. Captive flamingos that are not given pink shrimps to eat lose their bright colour and become a rather bland off-white. By contrast, iridescent colours such as the blue of a Kingfisher's back are structural, the result of light refracting from the feathers in a particular way, rather than of any actual colour being present in the feathers. Iridescent colours change appearance with light direction, and in isolation the feathers are colourless.

A Robin's colours are produced by two kinds of melanin pigment – eumelanin, which gives blackish and dark brown tones, and phaeomelanin, which gives red, orange and red-brown tones. Melanins are not derived from a particular dietary component, and their presence and intensity are determined by genetics.

Albinism, leucism and melanism

Above: This Robin's aberrant plumage illustrates how both types of melanin are needed to produce typical Robin colours. Without them their brown upperside is a ghostly grey and their red breast is almost pure white.

Certain genetic mutations can cause abnormal melanin deposition in the feathers, and result in some very peculiar-looking birds. The best-known colour mutation in birds, although by no means the most common, is albinism. Albino birds have no melanin at all so their feathers are white, and their bare parts (bill, legs and eyes) are pink because their blood shows through the unpigmented skin/retina. It is very rare to observe an adult albino bird in the wild, because albino birds' survival rates after fledging are very low. Not only do they stand out to predators, but the lack of melanin in their eyes makes them very light sensitive and may cause vision problems. It is easy to imagine how uncomfortable this would be for such a large-eyed bird as the Robin.

The condition leucism causes white plumage (it may affect the whole bird or patches of plumage, or sometimes just the odd feather), but does not affect the bare parts. Leucistic birds have normal vision and stand a much higher chance of survival than albino birds, especially in garden environments where there are few natural predators. There are also conditions whereby melanin is reduced, producing a washed-out appearance, and others where only one of the two melanin types is reduced or missing. A lack of phaeomelanin only, for example, produces a 'Robin whitebreast'. While such birds may survive as well as their normal-looking siblings, they may be at a social disadvantage because the red breast is an important signal in Robin communication, between rivals and courting pairs alike.

Birds can also develop with an excess of melanin. This condition is called melanism and is very rare – the degree of excess melanin varies, but a Robin with marked melanism would be entirely sooty-black. A wholly black or wholly white Robin may initially be difficult to recognise as a Robin, but its distinctive shape, big eyes and way of moving should make it identifiable.

Below: This female Robin 'whitebreast', lacking the Robin's most defining feature, is still deemed a good partner by a normal-plumaged male.

General-purpose equipment

To survive and thrive in today's fast-changing world, a bird (or indeed any organism) is best off having many strings to its bow. Species that are highly specialised are vulnerable, and tend to be first to decline or vanish when sweeping ecological changes take place in their habitats. Those with more versatile ways are insulated, at least to some extent, by their ability to use much broader ecological niches. Robins are true generalists – their body shape and in particular their bill shape is very 'generic bird', and this serves them well in the wide range of habitats that they can exploit.

Birds' bill shapes give an immediate idea of what their preferred diet is – for example, those of birds of prey are hooked for tearing meat, fish catchers have long, hook-tipped or serrated bills for gripping their slippery prey,

Below: The Robin's bill is not too thick and not too thin, so can handle soft invertebrates and also some tougher plant matter.

Above left: The Red Kite is a scavenger and sometime-predator, with a hooked bill for ripping up carcasses.

Top: Like most insectivorous birds in the British Isles, the thin-billed Willow Warbler migrates rather than becoming vegetarian in winter.

Above: Tree Sparrows become partly insectivorous in the breeding season, but their solid bills are best suited to a winter diet of seeds.

Left: The Dunlin finds most of its food by probing in soft mud, and has a proportionately long bill for this purpose.

while wading birds often have long, down-curved bills for probing soft ground. Among small birds, insectivores such as warblers tend to have slim, fine bills, ideal for picking their prey from leaf surfaces, bark crevices and so on, while seed crackers like finches and sparrows have stouter, heavy-based, conical bills. The bill of the Robin is intermediate between these two shapes, fine enough to pick insects or spiders from smallish spaces, but strong enough to break down softer seeds and even some nuts. While the Robin's diet is biased towards invertebrate prey, its ability to also take seeds, berries and other plant matter allows it to remain with us all year round, while most of our strictly insectivorous species have to migrate south in winter, before their only food source all but disappears during the cold weather.

Below: For a very high-energy food like a suet block, it is worth expending a little effort.

Above: Robins are not especially acrobatic but can learn to cling to a mesh birdfeeder.

The Robin usually feeds on the ground, but also gleans insects from leaves and branches, and while it is nowhere near as agile as Blue and Great Tits, it can learn to hang from a mesh birdfeeder when needs must. Its large eyes give it superior low-light vision, so it can begin its foraging day a little earlier and knock off a little later than most other birds. Perhaps most importantly of all, it is an opportunist, always finding creative and often courageous new ways to find food, and this flexibility of behaviour is as valuable a survival tool as its general-purpose anatomy.

Relatives and Namesakes

The world is full of robins – you only need to look at a list of all described bird species to see that. There are scrub-robins, magpie-robins and fly-robins. New Guinea has Ashy Robins and Smoky Robins, Australia has Yellow and Pink Robins, Africa has ground-robins, Siberia has blue robins and America has... American Robins. It would seem that our own Robin has many relatives. However, many of these birds are robin in name only and are not closely related to the original Robin, just named after it because of some passing resemblance – particularly the possession of a red breast. Our Robin *does* have closer relatives, but piecing together its family tree has proved a tricky task, one that is still going on today.

A genus of one

The celebrated Swedish botanist and 'father of taxonomy' Carl Linnaeus made the first serious attempt to classify all living things in a binomial ('two-name') system. Each distinct kind of organism was given a unique species name, and related species were grouped together in the same genus (plural genera). Linnaeus's bird list, at 574 species was around 5 per cent of those we know of today, did include the Robin, and the binomial name he gave it was *Motacilla rubecula*. Alongside it in the genus name *Motacilla* were various other small, insect-eating birds, including a number of European wagtails and warblers, and a couple of oddities such as a honeycreeper from Mexico and a tanager from Brazil.

That was in the mid-18th century. Over the next few decades it was recognised that many of Linnaeus's

Above: The name 'robin' has been given to many other bird species around the world: some very similar to actual Robins, others very different.

Opposite: Successful and unique – the latest DNA research suggests the Robin is the only species in its genus.

Above: The genus *Motacilla*, as conceived by Linnaeus, contained many small birds including the Robin. Now it is home only to the wagtails, like this Pied Wagtail.

placements were erroneous, and the Robin along with many other species was reassigned to a different genus, leaving *Motacilla* to the wagtails. The Robin's new genus, *Erithacus*, was created by Georges Cuvier, a French naturalist, in 1800. The Robin hung on to its species name, and ever since has been *Erithacus rubecula*. Two more species, the Japanese Robin (*E. akahige*) and the Ryukyu Robin (*E. komadori*), were later added to *Erithacus*.

Just as related species are grouped together in genera, so related genera are grouped together in subfamilies, and subfamilies into families. The Robin's subfamily is Saxicolinae, and it and the other members of that subfamily are known as chats. These are small, thrush-like birds that include redstarts, wheatears and stonechats, as well as the *Erithacus* robins. They were for many years considered a part of the family Turdidae, a large group that also contained the true thrushes. The resemblance between the smaller chats and the larger true thrushes is clear – all are bouncy, boldly patterned and sweet-voiced birds that tend to feed on the ground and eat mainly invertebrates. However, in nature resemblance does not always signify relatedness. Even the most careful analysis of the smallest anatomical details could sometimes lead a taxonomist down the wrong path.

Above: The Northern Wheatear is a fairly close relative of the Robin, and the similarity of this female is evident not in its plumage but in its shape, stance, and way of moving.

Above: True thrushes like this juvenile Blackbird do have a similar stance and bouncy character to the Robin, but DNA evidence shows they are not closely related.

The advent of gene sequencing studies in the late 20th century revolutionised the science of taxonomy. It became possible to compare two species at the DNA level, gene by gene, and reveal the exact extent of their differences (and even infer how many millions of years ago it was in their evolutionary history that a single ancestor of both species had lived). Gene sequencing work on the family Turdidae revealed that in fact the chats belonged in an entirely different family. So the Robin and its relatives had to be shifted to a different drawer of life's filing cabinet, as the subfamily Saxicolinae became a part of the family Muscicapidae, or Old World flycatchers. Evolution had taken the chats in a different direction from their more specialised fly-catching cousins, producing superficial similarities to the unrelated thrushes, but their genes revealed the truth of their history.

The taxomonic story is still not over for the Robin. Genetic work on the other two *Erithacus* species indicates that they may be in the wrong genus, as they show more similarity to the nightingales (genus *Luscinia*), or may they actually be best placed in a brand new genus. It thus appears that the Robin languishes alone in *Erithacus*, with no really close relatives anywhere in the world.

I enjoy our little chats

In Great Britain there are six breeding species of chat besides the Robin. They may belong to different genera, but all are in the same subfamily as our Robin and the (sub) familial resemblance is strong. They are round, leggy and upright little birds with pleasant (or at least distinctive) songs, and most are quite colourful.

Above: The Black Redstart looks rather like a monochrome Robin (apart from its red tail).

Redstarts The Redstart, a bird of deciduous woodland, has a particularly interesting link with the Robin. It is a summer visitor to Europe, and before the migratory process was understood the venerable Greek philosopher Aristotle declared that Redstarts transmogrified into Robins in autumn and did not return to their Redstart form until the following spring. Male Redstarts are stunningly decked out in red, blue-grey, black and white, while females are plain milky-tea brown, but both sexes have red tails, which they constantly shiver. This trait is shared by their coastal and urban cousin the Black Redstart, in which females are smoky-grey and males smoky-black. Although very rare in Britain, the Black Redstart is a ubiquitous town and village bird across many other parts of Europe, giving its curious gravelly song in numerous settings where a visitor from Britain might expect to hear the more tuneful voice of a Robin instead.

Stonechats and Whinchats Head to open countryside and you might find Stonechats or Whinchats, or both these birds. With a penchant for sitting pertly on the tallest, most exposed bush-tops and fence posts, these dumpy and perky little birds are much loved by birdwatchers. In both species the males are colourful and the females a little drabber. Whinchats migrate to Africa in winter, while Stonechats move from the heaths to more forgiving coastal habitats.

Above: Stonechats are open-country birds and, like Robins, they do not migrate.

Northern Wheatear There are many wheatear species in Europe, but only the Northern Wheatear breeds in the British Isles. It is a summer visitor and often the first harbinger of spring, appearing overnight on grassy south-coast headlands in small flocks. It breeds in open habitat and is a conspicuous bird, with black, white, peach and grey plumage, and a rather showy personality.

Nightingale The Nightingale, the final British chat, could scarcely be more different from the Northern Wheatear in both appearance and habits. This legendary singer skulks in thick, tangly woodland – you stand much more chance of enjoying its matchless music than of clapping eyes on it. If you do manage a look, you will probably be struck by its Robin-ish appearance. Imagine a slightly too big, slightly too long Robin, which has no red breast but is plain unmarked brown all over with a reddish tail, and that is a Nightingale.

The Japanese and Ryuku Robins placed (formerly) along with the Robin in the genus *Erithacus* both only breed in Japan, although the Japanese Robin (right) overwinters in south-east China. It looks superficially like our Robin, with a vivid orange head and chest. The Ryuku Robin is a rare and declining bird found only on a few islands in the Nansei Shoto archipelago. It is a striking bird, orange above and black below.

There are dozens more chat species across Africa, Europe and Asia, a fair few of them with 'robin' incorporated somewhere in their full names. The magpie-robins have pied plumage and spectacular long tails like their first namesakes, and beautiful songs like their second. The rock-thrushes are stocky, often brilliantly colourful birds of mountain ranges. The akalats are shy and rather mysterious birds of deep forest. No chats are found in Australasia or the Americas (except the occasional hopelessly lost migrant), but these countries are home to 'robins' of a different kind.

Below: Indian Robin.

The red, red Robin

The American Robin is a lot bigger and more colourful than its British namesake, but you could say that the same is true of North America itself. This common and handsome North American thrush with a red breast and brown back reminded early settlers of the Robins back home, so it was named the American Robin. The species is known simply as the Robin on its home turf, and is as much loved as our own Robin is in here. In view of the discovery that the true thrushes are not closely related to the chats, a renaming might seem appropriate, but no sensible taxonomist would try to convince North Americans that it would be a good idea to rename their robin Red-breasted Thrush or something similar. At least its scientific name (*Turdus migratorius*) reveals its true relationships.

The American Robin is a handsome bird, in shape and mannerisms reminiscent of a Blackbird or other typical true thrush. Its brick-red frontage extends to cover its belly but not its face, which is blackish with striking white 'spectacles' and throat-streaks. Females are a little duller and paler than males, and juveniles have pale, very spotty fronts. A common and widespread species, it is migratory in Canada and the most northerly US states, and the

Below: The blue, blue eggs of the red, red robin.

occasional migrant has actually wandered the width of the Atlantic to end up in the British Isles. However, across most of the US it is resident. Like our Robin, it is a bird of woodland and gardens.

Although the Robin of the Brits and the Robin of the Americans are quite unalike, confusion still manages to occur, one way or another. The film *Mary Poppins*, supposedly set in Edwardian London, features appearances by puppet Robins during the song 'A Spoonful of Sugar'; as well as being highly unconvincing, the puppets are clearly modelled on the American Robin rather than the British one. The phrase 'Robin's Egg Blue', denoting a colour, is widely used. As a consequence a British person discovering a British Robin's nest may be baffled to see that the eggs are not blue at all, but pale with reddish speckles – it is the American Robin that lays the blue eggs. Moreover, the song 'When the Red Red Robin (Comes Bob, Bob, Bobbin' Along)', from 1926, refers to the American Robin, but that did not stop the London football club Charlton Athletic (the Red Robins) adopting the song as its home-ground anthem.

Robins down under

The birdlife of Australasia is very unlike that found elsewhere on earth. The bird family Petroicidae, for example, has proved extraordinarily difficult to classify. However, British visitors would certainly notice a strong similarity between the European Robin and certain members of this (quite unrelated) family. Some petroicids have a distinctly Robin-like roundness, large-headed outline, perky stance, Robin-like body proportions and Robin-like habits. A few even have red breasts. Collectively, they are called the Australasian robins.

The family is a large one, with more than 40 species across about 14 genera. Within the genus *Petroica*, males of several species are particularly colourful and are known as the red robins. Among them are the beautifully rosy-breasted Rose Robin, the eye-poppingly orange-breasted Flame Robin and the frankly preposterous Pink Robin with its iridescent magenta underside. This

Below: The Flame Robin of Australia is perky and red-breasted just like our Robin. This is a female – males are even redder.

group also includes the much less striking but much more famous Chatham Island Robin, an endangered species restricted to the Chatham Islands, off New Zealand. With a total population of about 250 birds, its future looks shaky to say the least, but it has already come back from the brink of extinction due to an intensive breeding programme initiated when numbers were down to just five. Of these, only one was a fertile female, but she (almost) single-handedly saved her species by producing numerous clutches of eggs (when she was already much older than most songbirds are when they die), which biologists placed under foster parents of a different but closely related species.

The Robin's influence on humanity has crossed continents, as is evident by the many far-flung bird species that bear its name. It is not just birds that are called robins, either. Robin as a first name for both boys and girls is very popular in English-speaking countries worldwide – although originally the name came before the bird; it is often parents with a fondness for birds and nature who select it for their children today. As for the real, original Robin itself, this humble (well, actually, not that humble) garden bird has been the subject of a vast amount of scientific research, including some of the greatest classic behavioural studies of wild animals ever conducted, and as a result we know more about its unique way of life than we do about practically any other British bird.

Above: The spectacular Pink Robin breeds in wet woodland in southern Australia and Tasmania.

Below: The male Scarlet Robin is a strikingly beautiful bird. Unlike our Robins, most of the Australasian 'robins' show marked sexual dimorphism.

Anatomy and Adaptations

Although there are nearly twice as many bird species on Earth as there are mammal species, mammals show far more variety in body shape, so much so that the most extreme forms – whales and bats – were not even recognised as mammals by the earliest scientists. This seems to suggest that bird-kind hit on a winning body-shape formula early in its evolutionary history, with relatively minor 'tweaks' producing birds capable of thriving in a huge range of habitats. The Robin is very much a 'standard bird' in appearance, but like all species it carries an array of adaptations that allow it to successfully pursue a particular way of life.

A terrible legacy

One of the most difficult-to-swallow scientific revelations of recent years is the consensus that modern birds are dinosaurs. That is not to say that they evolved from dinosaurs – they literally are dinosaurs, being a direct evolutionary offshoot of a particular subgroup of dinosaurs called the theropods. These dinosaurs hold a prominent place in human imagination, including as they do the mighty *Tyrannosaurus rex*, as well as the *Velociraptor* species so memorably (if not accurately) portrayed in the film *Jurassic Park*. Since that film's release in 1993, palaeontologists have established that the majority of theropod dinosaurs, including *T. rex* (probably) and the velociraptors (definitely) had at least some feathers – and some even earlier dinosaurs did as well. In the case of velociraptors, there were some large wing feathers, evidenced by 'quill knobs' on the bones, showing where big feathers anchored. Modern reconstructions of velociraptors look very similar to those of *Archaeopteryx*, the 'first true bird'.

Above: You won't find the Robin on any list of the world's fastest or most agile avian flyers, but it is still a highly skilled aeronaut.

Opposite: An early bird, *Archaeopteryx* was around some 150 million years ago. It had well-developed feathers, but retained ancestral traits like a bony tail, toothed jaws, and claws on the 'fingers' of its wings.

Right: More than 100 million years more recent than *Archaeopteryx*, this fossil bird shows a modern avian skeletal structure with no teeth or tail-bones.

The evolution of the feather from its precursor, the reptilian scale, was a key event in the history of birdlife. The first soft, downy feathers allowed a dinosaur to trap heat and thus control its body temperature, and in due course larger, strong but extremely lightweight flight feathers permitted gliding and eventually flight. Today, birds' feathers really are the secret of their success, allowing full self-thermoregulation, highly effective waterproofing, a huge range of flight styles, a canvas for incredibly effective camouflages, and a means of communication via the display of particular colours or patterns.

The bird blueprint

The evolution of flight, and all the opportunities this opened up, meant that a lighter body (relative to body volume) became a big advantage for early birds. Weight savings were made in several ways – some bones became pneumatised (full of air holes); the bones in the tail were lost, with long, sturdy feathers providing the necessary rigidity instead, and teeth were discarded altogether. The avian bill or beak, with the jawbones covered by strong but light keratin, has diversified into an array of shapes, adapted for tasks varying from hammering holes in solid wood to stitching leaves together with plant fibres to make a nest.

Above: Feathers, once thought to be a defining trait of birds, are now known to have been present in a wide range of dinosaur species, especially the theropods.

While birds' forelimbs became the means for flight (and/or strong underwater swimming in a few cases), the hindlimbs also diversified in a range of ways, with webbing for swimming, hooked talons for clutching prey, or flexible strong toes for climbing and manipulating objects. Some species retained their dinosaur-like, long, strong legs and broad, tough feet, for running faster than anything else on two legs.

In the Robin we see a fairly typical avian anatomy. The body is fully feathered, with a down layer for insulation overlaid with smooth contour feathers to hold or release heat as required. The long and strong flight feathers are

Above: A Robin's plumage is a canvas to show off colour, and a waterproof, insulating 'coat' to keep it warm and dry.

arranged to form a medium-length, somewhat rounded wing, which provides a compromise between power and manoeuvrability. The wing bones contain air spaces that save weight, but at the same time are formed from relatively dense tissue, and their shape is adapted for maximum strength to stand up to the stresses of flight. The legs are relatively long, allowing the bird to hop strongly and stand tall to obtain the best view (helpful for a species that mostly forages on the ground). Like most birds, the Robin has four toes per foot, with three pointing forwards and the fourth backwards – this gives good stability on the ground and allows a strong grip when perched on a branch. The legs and feet are unfeathered, but the Robin can keep them warm when necessary by adopting a crouched posture when perched so that the body plumage covers the legs and feet.

Internal workings

The digestive equipment of a Robin begins with the mouth. Lacking teeth, the Robin swallows its food whole, whether that food is an earthworm, a berry or a suet pellet from a bird table. Everything goes into its crop, a sort of storage area where quite a substantial volume of food can be stashed, but where little digestion takes place. From the crop food proceeds to the proventriculus, where acid and enzymes begin the digestive process in earnest, and then into the gizzard. Here it is subjected to intensive muscular pounding and dousing with digestive enzymes.

Anything very tough and resistant to enzyme action progresses no further. Such material is compressed into a pellet, which the bird ejects from its mouth. We tend to think of only birds of prey as pellet producers, but in fact many species – including Robins – cough up pellets when necessary, although the frequency at which a bird does this depends on its diet. A Robin's pellet might contain beetle wing-cases, hard seeds from inside a berry, the tough parts of centipedes and even the shells of small snails. However, if a Robin eats a quantity of entirely soft food, no pellet will result.

Below: An earthworm is a high-value food item, large and entirely digestible, so worth the extra effort to obtain.

Further digestion and absorption of nutrients and water goes on through the bird's intestines, and whatever remains is excreted via the cloaca. At the same time, waste products from the blood-filtering action of the kidneys are added. This is the bird's version of urine, but it is white and often semi-solid rather than a clear fluid as in mammals.

Like mammals, birds breathe to obtain the oxygen they need to power the cellular process that releases energy for bodily functions, and to get rid of the carbon dioxide that is a waste product from this chemical reaction. As well as a pair of lungs, where the oxygen/carbon dioxide exchange takes place, they have a system of large, connected air sacs in their bodies, and because of the arrangement of these (along with the air spaces in the bones), fresh air constantly flows into the lungs while 'old

Below: When a Robin eats prey that has indigestibly hard parts, those bits are later ejected from the mouth in the form of a pellet.

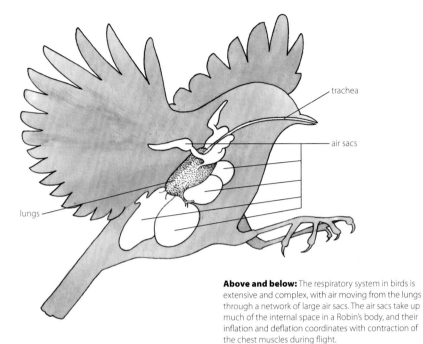

Above and below: The respiratory system in birds is extensive and complex, with air moving from the lungs through a network of large air sacs. The air sacs take up much of the internal space in a Robin's body, and their inflation and deflation coordinates with contraction of the chest muscles during flight.

air' bypasses the lungs on its way out. This makes birds' lungs much more efficient at gas exchange than mammals' lungs, and also makes birds into what are in effect feathered balloons. When a bird is on the wing, inflation and deflation of the air sacs works in synchrony with the muscular action of flight.

The avian heart and circulatory system is much more similar to that of mammals, with a four-chambered heart pumping oxygenated blood around the body via arteries, and deoxygenated blood returning to the heart along veins via the lungs, where it is 'reloaded' with oxygen. However, a Robin's heart is proportionately much larger and stronger than that of a mouse or other similar-sized (and non-flying) mammal.

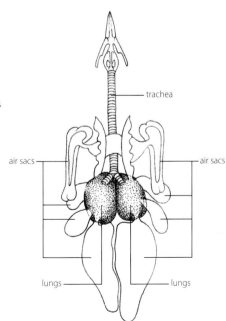

Senses and sensibleness

As already mentioned, Robins have large eyes relative to their size. A large eye means a large light-gathering retina, and this allows Robins to forage successfully where ambient light is low – at dawn, dusk and in very shady conditions. Like most other birds, Robins have four types of colour-detecting cone cell in their retinas, and can see ultraviolet light as well as detect more subtle colour differences than human eyes can manage. The other key sense is hearing, which is highly developed in Robins as a means of finding prey, detecting danger and managing communications with other Robins. A Robin's sense of smell is probably not very acute. However, recent research in this area, including genetic investigation, is uncovering increasing amounts of evidence that birds of many species have a more developed sense of smell than had previously been thought. Like many small insect-eating birds, Robins have stiff modified feathers called rictal bristles around their mouths, and these probably have an important role in sensing prey by touch.

Below: With their large eyes, Robins can remain active well into dusk, and even all night if there is strong moonlight.

Magnetoreception

The mysterious 'sixth sense' of magnetoreception, or the ability to sense magnetic fields, is observable in many birds and has been investigated in Robins. The ability is down to the presence of crystals of the iron oxide magnetite within receptor cells in the skin of their bills. These cells send messages to a particular part of the brain in response to magnetic fields, but this 'compass' system also requires a certain level of light to work properly. The combination of (seasonal) changes in light levels, and the Robin's ability to detect the Earth's magnetic field, allows it to orient itself correctly for migration. Some Robin populations do not migrate, but the presence of the magnetite-containing receptor cells has been demonstrated in birds as non-migratory as domestic chickens, so it is likely that magnetoreception has other functions besides a role in guiding migration.

The Robin's brain accounts for 3–4 per cent of its total body mass, which is about average for birds of its size (although both Great and Blue Tits have significantly bigger brains). In general, smaller animals have larger brains relative to body size, and clever animals have relatively larger brains than their slow-witted cousins, but brain size is just one factor that predicts intelligence – the details of brain anatomy are also important, and some brain structures that have a basic function in mammals have been found to deal with higher level processing in birds. Most scientists now agree that birds such as crows have intelligence levels on a par with brainy mammals

Above: The Robin uses visual input combined with information transmitted by the magnetite-containing nerve cells in its bill to navigate.

such as monkeys. The brain structure concerned with learning and performing song (the HVC or higher vocal centre, only found in songbirds) is particularly large and well developed in Robins.

Measuring intelligence in any animal is tricky, not least because it is difficult to agree on what intelligence actually is. A complex sequence of behaviours, such as a Robin pair finding a safe nest site, constructing a fit-for-purpose nest and raising a family, with all the coordinated activity that requires, seems pretty clever, but if it is all predetermined by instinct does it really qualify? Showing the ability to invent or learn new, innovative behaviours is universally taken as a sign of true intelligence, and here Robins excel, particularly in their ability to exploit human habits for their own gain.

Below: Is choosing to nest inside buildings an intelligent decision? In most cases it works out well for the Robins as predators are kept away.

However, there is also some evidence of a distinct lack of common sense in Robins under certain circumstances. When conducting his famous Robin studies in the 1930s around Dartington Hall in Devon, biologist David Lack used traps to catch subjects for colour ringing. He found it very easy to catch Robins – they would readily enter a trap within minutes of it being placed – but other small birds were warier of the traps. Moreover, the same individual Robins would get caught again and again, while other species were much less likely to be retrapped. Perhaps the trapping process was not very stressful for the Robins, or maybe the same insatiable curiosity that helps the birds to use hanging birdfeeders, go into garden sheds to nest in flower pots, and follow gardeners in order to obtain worms also results in them being prone to getting themselves into trouble.

Below: Watching gardeners digging over the soil to reveal worms is a trick most garden Robins seem to learn quickly.

Territoriality and Song

One of the reasons why we have a particular fondness for Robins is that we usually only see one in our garden at a time, so we tend to assume we are seeing the same Robin every day. In the short term at least, this is probably true, but people who claim to have been feeding the same Robin for 10 years are likely to be wrong. Territory is everything for Robins, and if a top-quality territory becomes vacant when its owner dies, a new Robin will move in so quickly that the change will be unnoticeable to all but the keenest observer. One of the other most appealing Robin traits, its lovely song, is the tool it uses to declare and defend its territory.

A place of my own

Most birds have some kind of territory that they defend from others of their own kind. This might only extend a few centimetres beyond the nest, as in highly colonial seabirds, or in the case of large birds of prey may cover many square kilometres. Some birds, including most of our familiar garden visitors, defend a territory when breeding, but in winter they lose their territorial ways and forage together in flocks. Robins, however, guard a territory all year round and are notorious for being exceptionally intolerant of intruders.

What does a Robin need from its territory? The key thing is a good and reliable food source, in the form of habitat for insects and other invertebrates – a good mix of reasonably dense native vegetation and leaf litter of the sort found in a lowland deciduous wood is ideal. The food source also needs to be safely accessible, and the territory has to have places in which to drink, bathe, rest and take cover from danger. Additionally, in the breeding season it needs to provide a safe nest site (or, even better, a choice of nest sites). It needs to hold

Opposite: Song is key to Robin interaction, at all times of year and between both sexes.

Above: Robins often sing from prominent perches, giving the visual signal of their red breasts as well as the auditory message of the song.

all of these resources, but within an area that is small enough to enable the Robin to effectively defend it without having to spend all of its time rushing from one end of the territory to the other to guard the borders – its song needs to be heard by potential intruders from all sides. Overall, quality of habitat is more important than territory size when it comes to breeding success.

When breeding, a pair of Robins necessarily shares a territory. However, in winter the pairs usually split up and males and females hold separate territories – males are much more likely than females to remain in the territory that they used for breeding. In winter, although territorial behaviour still occurs, it is reduced and, particularly in cold weather, Robins can become quite tolerant of each other. Defending a territory costs a lot of energy, and when the temperature is low and food is difficult to find, chasing off rivals is simply not worthwhile. However, at a very concentrated source of food (say, a garden bird table laden with suet pellets or mealworms), a Robin may quite fiercely chase away not just other Robins but all other small birds.

Territory size and stability

Much of what we know today about Robin behaviour is derived from a study of Robins carried out by David Lack. By trapping Robins and marking them with unique arrangements of coloured leg rings, he and his students were able to re-identify individual birds while observing them engaging in their natural behaviour. The birds' territories were mapped, and the stability of the territories over subsequent months was recorded.

Territory size, whether of a single bird or a pair, tended to average at about 0.61ha (1½ acres), although the local Robin population increased over the first two years of the study and some territories grew smaller as new birds squeezed in and borders were renegotiated. Whenever a bird went missing (presumed dead), its territory would quickly be invaded and occupied by others – Robins notice when one of ther neighbours has ceased to sing, and begin to make tentative incursions into the vacated territory, hoping to expand their own. There are also often young birds existing sneakily at other Robins' overlapping territory edges or in less desirable habitats nearby, watching, listening and waiting for the chance to secure a patch of their own.

Above: As juvenile Robins begin to moult to adult plumage, they must leave their parents' territory and look for a patch of their own.

Above: Uniquely marking individual Robins with colour rings allows researchers to keep track of changes in territory ownership and boundaries.

David Lack found that in his study area most winter territories were held by males. So where do the females go? Recent research has found that females greatly outnumber males in southern Iberia in winter. This suggests that female Robins are more likely to migrate south in winter, and they (as well as young males in their first winter, and smaller-than-average males) are more likely to be found in less optimal habitat, probably because they are excluded from the best territories by larger and older males. However, some females do successfully hold winter territories in prime habitat, and defend them in the same way as males – with song.

An authoritative voice

The Robin's song is easy to recognise. Sweet, varied and high-pitched, with fast twitters and slower phrases, it is most notable for the wistful, almost melancholy form of its melodies. This aspect is particularly pronounced in autumn and early-winter song, when the birds are less strongly driven to sing and thus deliver their tunes rather less forcefully than they do in spring and summer. Robins sing most around dawn and dusk, as sound travels better at these times than during the middle of the day. This atmospheric quirk is down to dawn and dusk air being cooler at ground level than above, which bends sound waves downwards. In urban environments there is also much less ambient sound at these times, particularly at dawn. Robins do sing on and off throughout the day, and also sometimes at night.

The discovery in the 1920s by Irish researcher J. P. Birkitt that female Robins sing was a surprise for the ornithological community. Before that it was generally presumed that song was strictly for the males. It has since been discovered that females of many other species do sometimes sing, and in at least one case actually use song to attract a mate, as well as in territorial defence.

Above: Robin song is at its most intense on spring mornings, but can be heard at any time of year, and any time of day (or night).

Female song in Robins is heard in autumn and winter, as a territorial defence signal, but not in spring. Studies have shown that female Robins show raised testosterone levels outside the breeding season, and if treated with testosterone in spring they sing at this time of year as well. The loudest and most enthusiastic Robin song of all comes from males in early spring, when seeking to attract a mate as well as defend the territory, and this intense singing phase is also triggered by raised testosterone.

As mentioned earlier, Robins and other songbirds have a distinct brain structure (the HVC) which handles the learning and production of song, via neural links to the ears and the sound-perceiving parts of the brain. In Robins this structure is twice as large (by volume) as it is in many similar-sized songbirds such as the Chaffinch, Blue Tit and Blackcap, and this correlates with the much more varied songs of Robins compared with those of the other species. Young Robins need exposure to adult Robins' songs to learn how to 'sing like a Robin'; if hand-reared baby Robins are isolated from Robin song and exposed to other species' songs they develop songs resembling those instead. David Lack noticed with his captive-bred Robins that there was an increase in the adult male Robins' songs at around the time when their broods of chicks fledged; this may help the learning process (although the chicks do not make their own first attempts to sing until some weeks later).

The winter nightingale

Night song is rather common in Robins, leading many inexperienced listeners to believe that they are hearing a Nightingale. It is fair to say that a bird heard singing at night in winter in a town is not going to be a Nightingale, as that species is strictly rural and also a summer visitor (although according to a Downing Street anecdote, these facts were not sufficient to convince Prime Minister Margaret Thatcher of the true identity of the night-singing bird that she heard one winter). Robins are most likely to sing at night in areas with all-night street lighting, and they begin their dawn song earlier in urban areas than do their country cousins. The duration of the song and the onset of dusk singing is not affected, however. The increased light levels in towns have been shown to extend foraging time as well as singing time, and urban Robins are very likely to have different sleep patterns from rural birds. Whether the extra hours of night-time activity are actually detrimental to Robins in towns has not yet been established.

Left: Night-singing Robins are often misidentifed as Nightingales. Sweet as Robin song is, the real deal is unmistakable in its richness and complexity.

This is war

Above: In a state of high agitation, a Robin sings while posturing to show off the red breast to its best advantage and intimidate a nearby rival.

Song is a Robin's way of communicating to a rival, 'Stay out of my territory, or else'. Often it is enough of a deterrent on its own. However, if it is not, then things escalate, via aggressive posturing, chasing and, in the most extreme encounters, full-on brawling – which can result in serious injury or even death. That is how important territorial defence is to this little bird.

When defending its territory, a Robin confronts any

other Robin of either sex that breaches the borders (apart from its own mate or a prospective mate). When a Robin is new to its territory, its neighbours will be particularly vigilant in reminding the newcomer to keep to its own patch. This was demonstrated in an experiment whereby territorial males were trapped and briefly removed from their territories, on being returned they sang and displayed with exceptional vigour, apparently aware that trespassing could have occurred in their absence. Trespass is most likely to be tolerated in harsh winter weather, and also during the moult in late summer, when food is abundant, there are no dependent chicks to worry about and no Robin feels at its best. Most trespassers are pushing their luck and avoid all confrontation with the territory owner, but some may be intent on a takeover bid, and this is when things escalate.

Aggressive displays between Robins are focused around the red breast-patch. When face to face with a rival a Robin tilts its head up, sticks its chest out

Seeing red

The Robin's red breast-patch is used as a signal in aggressive displays. Studies using stuffed Robins and bits of Robin parts have demonstrated very graphically the overriding power of this signal. Robins almost invariably attack a stuffed specimen of an adult Robin placed in their territories, and do so with sustained ferocity that eventually leads to the total destruction of the specimen. However, if the stuffed bird's breast-patch is dyed brown, a wild Robin ignores it completely. Even more strikingly, if just a bundle of red breast feathers is placed in a Robin's territory, the Robin will attack the clump with just as much vigour as if it was an entire bird.

Right: The more a young Robin's breast-patch develops, the more likely other Robins are to react to it with aggression – even its own parents.

and cocks its tail. If the rival is above or below the defender, the posture is adjusted accordingly. Holding this exaggerated posture, the Robin slowly sways from side to side, an action very different from its usual brisk movements and certain to catch the target's attention, especially as it is accompanied by particularly shrill and emphatic song. The defender next flies at its rival and this is really the last-chance saloon – either the intruder flees with the territory holder chasing it out of the territory, or things get physical.

Fighting is rare, and ferocious. The combatants clutch at each other's feet and direct a furious volley of pecks at each other's heads. Feathers literally fly. If the battle began in a bush or tree, the birds often tumble down to roll and wrestle on the ground, with much angry squealing. They are completely absorbed in their activity and will do this at the feet of an observer (or, if they are unlucky, a predator). The bird coming off worse eventually tries to escape, but sometimes these are fights to the death; one (or both) birds may be so injured from the fight that it dies not long afterwards.

Above: If an intruder to a territory is not deterred by the territory holder's song, intense face-to-face posturing will follow.

Diet and Feeding Behaviour

As already mentioned on page 14, the Robin is not a specialist in terms of the types of food it eats and how it finds the food. Its adaptability in this respect is key to its success, as is its ability to forge close ties with the humans that share its space (as long as they can provide it with food!). Most other birds of its size have special skills – for example, tits dangle from twigs to pick caterpillars from tree leaves, thrushes listen for worms moving in the soil, Treecreepers climb tree trunks and probe bark crevices, and flycatchers deftly capture insects in flight. However, Robins can do all of these things too – albeit somewhat cack-handedly compared to the experts – so the range of feeding opportunities available to them is huge.

Basic rations

If you spend some time watching a Robin during the spring and summer as it searches for (natural rather than bird table) food, you will see that it spends most of its time on the ground, hopping along with frequent pauses to look around. It searches both grassy and bare ground (but not so much in very thick ground-level vegetation), and may turn over leaves or loose soil to see what is underneath. What it picks up is usually live prey – a beetle, woodlouse, spider, millipede or worm. It may take considerable numbers of ants, and certainly exploits a stash of insect larvae if it finds them.

Above: It's much easier to find worms and other soil invertebrates if a helpful human digs the ground over first.

Through autumn and winter the natural diet changes, sometimes dramatically, with fruits and seeds becoming much more important. A study of Robins wintering in Spain found that the plant component of the diet

Opposite: The British habit of feeding garden birds benefits many species, including the versatile and fast-learning Robin.

Above: Robins can forage successfully in fairly short grass, so appreciate a garden lawn.

averaged as high as 77 per cent by volume, with the proportion of animal matter gradually increasing during late winter and into early spring. This reflects the change in abundance of insects and other invertebrates, but it also makes good nutritional sense for the Robin to take more plant matter in autumn and winter than it does in spring and summer. Seeds in particular can have a very high fat content, and fat is extremely important for small birds, particularly when they are preparing for the rigours of migration and having to survive winter's long, cold nights. No other natural food provides as much energy in as dense a form as seeds do.

In gardens, Robins come to bird tables and feed on almost everything on offer, from peanuts and sunflower seeds to suet in any form, and kitchen scraps like overripe fruit, bits of cheese, bacon rinds and even breadcrumbs (although the latter contain very little useful nutrition for a bird). However, when rearing young they really do need the easy-access protein provided by soft-bodied invertebrates, so if you want to keep your Robins happy all year round, a wildlife-friendly garden with space for creepy-crawlies is the way to go.

Robins and mealworms

The mealworm is the larval stage of a medium-sized black beetle (*Tenebrio molitor*). This beetle lays its eggs in flour and grain stores, so it can be a pest, but it has also acquired great popularity as an easy-to-rear live food for various exotic pet birds and reptiles, and for wild birds. Anyone wanting to make friends with their garden Robin should invest in some mealworms, for there is no food more irresistible to the birds than these light-brown, legless grubs.

Above: Mealworms, like other soft-bodied invertebrates, are prime Robin food as they are full of protein and fat, and there is no waste.

Rearing mealworms at home is quite easy. The RSPB provides detailed guidelines for would-be mealworm-farmers, as described below.

For a constant supply of mealworms, prepare a large, circular biscuit tin as follows:

1   Punch small holes in the lid of the tin for ventilation.
2   Place a layer of old hessian sacking in the bottom of the tin and sprinkle fairly thickly with bran. Place a slice or two of bread and raw potato on top, followed by another two layers of sacking/bran/bread/potato. Place a raw cabbage leaf on top if you like. Keep the tin at room temperature, not in hot sun.
3   Place 200–300 mealworms in the prepared tin. After a few weeks the mealworms will turn into creamy pupae, then into little black beetles.
4   The beetles will lay eggs, which will hatch into mealworms and so on. Crop as necessary. Replace the bread, potato and cabbage as necessary.
5   If you want to start new colonies, prepare another tin and transfer some bits of dry bread (these will carry beetle eggs) from the flourishing colony.

Above: A fisherman's box of wriggling maggots proves irresistible to an opportunistic Robin.

If a home mealworm farm is not for you, live mealworms can be bought from many bird-food retailers. A range of sizes may be on offer – consider mini-mealworms for during the breeding season, as these are suitable for small nestlings (the adults will collect the worms and take them to the nest). If all the wriggling is just too much for you, it is also possible to buy dead, dried mealworms, which you

Right: Winning a Robin's confidence begins with offering food from the hand – actually persuading it to perch on your hand takes a little more time.

Above: With mealworms on offer, most garden Robins will quickly become brave enough to feed from the hand.

can soak to restore their plumpness, although the birds will also eat them as they are.

You should be able to persuade your garden Robin to take mealworms from your hand with a little patience, especially once it has become accustomed to finding mealworms on offer in your garden. Start by finding a spot where you can comfortably sit for a while, and try placing some mealworms in plain view, not far from where you are sitting. For example, if you have a garden bench you could sit at one end of it and place the mealworms on the opposite arm. Once the Robin is coming for these, offer some a bit closer, and work towards offering them on the palm of your opened hand. Some people have had success within hours. It helps if your Robin is already used to seeing you around the garden, although most Robins have enough innate curiosity, and a passionate fondness for mealworms, to become hand-tame quite quickly. Once the Robin has built a strong association between you and food, you may find that it will come into your house through an open door or window to demand its rations, so it is advisable not to encourage too much tameness if you have a pet cat.

New ways of foraging

Above: Fat blocks are very useful energy sources in the cold depths of winter.

Below: Robins generally prefer to feed from the ground or flat bird tables rather than hanging feeders.

Although the Robin is basically a ground feeder, it also looks for food in bushes and trees, and most of the berries it takes are picked from plants rather than collected on the ground. The feeding method known as hover-gleaning, which is perhaps best demonstrated in Britain by the tiny Goldcrest, involves hovering around leaves that could not be reached while perched, and picking off prey like aphids, scale insects and small spiders. Robins can do this after a fashion, and may be seen doing something similar at hanging birdfeeders that do not have horizontal perches (such as mesh cage-type feeders that are designed to be clung to rather than perched at); they seem to be a little more skilled at hovering than at clinging and dangling. Robins also adopted the Great and Blue Tit behaviour of stealing the top of the milk when doorstep deliveries of bottled, foil-capped milk were still common.

Robins may belong to the flycatcher family, but actually catching flies on the wing is not their forte. However, they can manage quite well with the slower and more clumsy flying insects, such as winged ants, plump alderflies and any freshly emerged insect that has yet to gain its full wing strength. One way in which a Robin can easily feast on flies is to glean squashed ones from the tyres and grilles of parked cars. This behaviour is common in the Pied Wagtail, but some enterprising town Robins have learned to do it too, using the hover-glean technique to access the grilles.

Gardeners have noticed for many years that the garden Robin is usually nearby when earth is being turned over. Hanging

around big animals to snap up the insects they disturb is something quite a few birds do – Yellow Wagtails are well known for following cows and horses around their pastures in Britain, and Cattle Egrets do this in many countries. For Robins in Europe, following herds of foraging Wild Boar often proves rewarding. The boars expose plenty of worms as they unearth plant roots for themselves. Wild Boars were hunted to extinction in Britain in the 17th century, but now they are back and on the increase, after reintroductions and escapes from farms throughout the late 20th century. Not all conservationists think this is a good thing, but Robins seem sure to benefit. Boars are especially good for breaking up snow-covered, frozen ground, so they could be life-savers for Robins in a prolonged cold snap – especially in late winter, when berries and seeds are becoming hard to find. The same goes for people walking in snowy countryside – if you notice a Robin watching as you walk, try kicking away a patch of snow to give the bird a chance to look for prey. Sometimes even groups of people sledging downhill in country parks will attract hungry Robins, which will flit over to investigate ground cleared of snow by a sled's rails. On rare occasions, Robins will even feed on carrion.

Above: Insects squashed in a car grille can be a valuable source of food for the urban Robin.

Below: Snow cover makes foraging difficult for Robins, and they will take advantage of any patches scraped clear by people or vehicles.

Breeding

Securing a territory and guarding it from all comers. Singing, challenging and fighting. Surviving the cold winter nights. Taking in enough food every day. None of it means a thing if a Robin does not go on to breed successfully. The breeding season takes up half of each year, and a pair of Robins could produce three broods of five chicks through spring and summer. The parents invest massive effort into each breeding attempt, overcoming long odds to ensure that their genes are passed on to the next generation.

Pairing up

When male Robins start to sing with renewed, testosterone-powered vigour, sometimes as early as mid-winter, the local female Robins take notice. They will already be aware of their neighbours and which males hold the best territories, and to them the song is an invitation to enter the territory. Often a female will be greeted with aggressive displays by a territory-holding male, and she may posture back in kind. However, over the next couple of hours (or, exceptionally, couple of days) this behaviour dies down and the pair settles together, although in the early days it is not unusual for the female to move on if she notices a more attractive prospect nearby. Males that lose their first mate are likely to attract a replacement sooner or later, as a certain proportion of females overwinter well south of their breeding areas.

Above: Courtship feeding, whereby the female solicits food from her mate, is a key part of courtship, pair-bonding and preparing to breed.

Opposite: A young Robin, just a few weeks out of the nest, will already be planning for its first breeding season the following spring by finding the best territory it can.

Robins sometimes re-form pairings of the previous breeding season. This occurs mainly when the male and female have spent their winter as next-door neighbours, although there are cases on record of ringed females that migrate well away from their breeding grounds, then return to the same patch and pair with their previous partner. There are advantages to sticking with the same mate – especially if the pair bred successfully in the past. However, it is the exception rather than the norm for previous pair bonds to re-form, mainly because of the Robin's typically short lifespan. While there is a tendency for females aged two years or older to go back to the territory where they bred previously, the territory is more important than the male that occupies it. If a female's previous year's mate has been displaced to a neighbouring territory, the female is more likely to take up with the new holder of the original territory than join her former mate in a new patch.

Very occasionally a male Robin can pair with two females at once. In one observed case of such polygyny, the two females had their own territories within the male's larger single territory. While this may mean double the amount of offspring for the male, dividing his efforts between two nests can be detrimental to all three birds.

Once a male has accepted a female into his territory, the pair occupies the same patch peacefully and the male's song becomes less ardent. There is not a great deal of interaction between the birds for several weeks, although the male tends to follow the female in a casual-looking way as she forages. Activity between the pair only begins to intensify when the female builds her nest, which usually happens in March, but may sometimes occur as early as February.

Below: Paired Robins may peacefully share a territory for weeks before breeding begins.

Nests and where to put them

Robins are notoriously adventurous in their choice of nest site. They look for some sort of hollow that offers both support and shelter, often among roots, within a thick stand of ivy or tucked into other dense vegetation. Sometimes it is placed inside the old nest of another (larger) bird. In gardens, however, the hollow is frequently not a natural one.

Above: Soft, springy moss is a key component of the outer cup of a Robin's nest.

Only the female Robin builds the nest. This is an open cup built on a foundation of moss and dead leaves, and lined with soft grasses, feathers and other cushiony material. In some cases it may have a partial roof for extra shelter. The builder is no slouch and she can complete her nest in a matter of hours, although more often the building is done in shifts of a few hours at a time over three or four days, while the rest of the time is spent feeding up to prepare her body for egg laying. This is also the time when the male and female's relationship moves on to the next level.

Nesting near humans

Robins readily use open-fronted (and occasionally hole-fronted) nestboxes, especially if they are sited on a wall or trunk that is well screened by foliage. If a shed or outbuilding is permanently accessible they may go into it and find a place to nest, perhaps on a shelf, in a flower pot or in the pocket of a forgotten coat. There are cases of Robins nesting under the wheel arches of cars left parked for a long time, inside opened bags of compost, and in one particularly inconvenient case, within the folds of an unmade bed in a room where the window was left open for just a few hours in the morning. The goal is to find a spot that is as safe as possible from predators, and many garden Robins seem (quite rightly) to recognise that nesting in close proximity to humans is an effective way of meeting this goal.

Courtship feeding and copulation

At this point in the breeding season you may observe a pair of Robins actively foraging quite close to each other. One bird keeps up a near-constant series of soft calls – this is the female. From time to time she stops still and begins to call more insistently. As her excitement reaches a crescendo she adopts a 'begging chick' posture, crouching and wing-shivering. The male appears at this point and quickly posts an item of food into her open mouth. Satisfied, she gulps down the worm or slug or whatever, and the pair resumes its previous activity. At times the female does not actively forage at all but waits,

Below: By feeding his mate, allowing her to rest and put on weight for egg laying, the male improves his own chances of successful breeding.

sitting still, calling quietly and watching her mate; she begins to beg when she sees that he has caught something sizeable and tasty.

The value of this courtship feeding is twofold. The male gets the opportunity to demonstrate his provisioning skills – if the female considers that he is lacking in this department it is not too late for her to cut her losses and seek a new partner. More importantly, however, extra food for the female helps her to improve her body condition and take on board the extra nutritional resources she needs for forming her eggs. Of course, she cannot complete this process without the male's contribution, and she solicits mating by adopting a still, hunched posture. Mating and courtship feeding might be seen within a short time of each other, but there does not seem to be any direct link between the two behaviours.

Below: This female Robin, close to egg laying, is waiting for a food delivery from her mate, reminding him of his duties by constantly calling to him.

Eggs

Above: Dunnocks and Robins build similarly sized cup nests, but Dunnocks' eggs are vivid blue.

Below: Robins have pale eggs with variable amounts of reddish speckling. This clutch of seven is unusual – most clutches are of four or five eggs.

A couple of days after successfully mating, the female is ready to lay her first egg. This usually takes place in the morning, and another egg is laid on each successive morning until the clutch is complete and incubation begins. In Britain most clutches are of five eggs, but six and four are also common, seven less frequent and anything outside this range very unusual. Elsewhere across the Robin's range, and to a lesser extent within Britain, the average clutch size varies somewhat, with the most northerly breeding birds laying more eggs, and southerly breeders laying fewer. This is a common trend in birds of all kinds. The larger northern clutches are counterbalanced by the fact that due to the shorter northern summers the Robins there tend to have two rather than three broods per year.

The eggs are about 2cm (¾in) long and whitish with a liberal dusting of reddish spots. Anyone seeing their first Robin nest and being surprised that the eggs are not 'Robin's Egg Blue' has been bamboozled by the popularity of 'robin' as a name for red-breasted birds elsewhere in the world – it is the American Robin that lays sky-blue eggs (see page 24). Due to the same misunderstanding, the nests of Dunnocks (which do lay blue eggs) are often presumed to be Robin nests, but in fact the Robin egg, although not very striking, is quite distinctive and not readily confused with other British bird eggs.

The Robin and the Cuckoo

Above: A Robin parent struggles to keep up with the voracious appetite of its adopted infant.

The only brood parasite of small songbirds in Europe is the Cuckoo, or Common Cuckoo, an elegant, long-tailed, hawk-like bird that arrives in Europe in mid-spring and by July is already on its way back to Africa, its offspring left behind in the nests of other birds. Dozens of species have been recorded as hosts for Cuckoo chicks, but by far the most popular in Britain are the Reed Warbler, Meadow Pipit and Dunnock. The Robin is in the top ten if not the top five, but the actual rate of parasitism for all hosts outside the top three is very low. Out of 12,917 Robin nests checked as part of the BTO's nest-record scheme between 1939 and 1982, fewer than 0.5 per cent contained Cuckoo eggs or chicks, and the rate is believed to have fallen since then, as Cuckoos have declined considerably. Robins nesting in typical gardens are almost certainly safe from Cuckoo activities. On the Continent things are rather different, and in eastern Europe the Robin is an important Cuckoo host.

Individual Cuckoos stick to one host species, and the distinct groups of Cuckoos are known as gentes, so there is a Robin gente as well as a Dunnock gente, Reed Warbler gente and so on. Each gente is genetically as well as behaviourally distinct from others, and the females lay eggs that are uncannily similar in appearance to their hosts' eggs. The female Cuckoo visits an unattended nest when the clutch is incomplete or when incubation has only just begun, removes one host egg and replaces it with her own. A minority of host pairs spot what has happened and abandon the nest, but most show no sign of noticing a problem.

After a couple of weeks the Cuckoo chick hatches and, in a grimly purposeful routine familiar from wildlife documentaries, gets rid of the remaining host eggs (or chicks) by manoeuvring them onto its back then hefting them out of the nest cup. The parents, apparently still unable to detect the deception even though it is far more obvious now (or perhaps they are just unable to resist the Cuckoo chick's large open mouth), rear the interloper as their own. If the youngster was reared by Robins, when it returns the following year it will seek out Robin nests if it is a female, and areas with many Robin territories if it is a male.

Incubation and hatching

Having built the nest and laid the eggs, the female Robin's next job is to incubate the eggs for 13 to 14 days. The male does not assist with this, but has the important role of keeping his mate well fed while she is incubating. Usually when she needs food she leaves the eggs and calls for him, but he does occasionally also feed her on the nest. He also alarm calls to warn her when danger is nearby.

Female Robins develop a brood patch to assist them with incubation. This is a bald area on the belly full of blood vessels, which transfers heat very effectively from skin to egg. The shed feathers provide extra cushioning and warmth in the nest cup, and the unshed feathers around the brood patch help hold the heat in. This is

Below: Hatchlings soon recover enough energy to shout for their first meal.

particularly important for the first clutch of the year – Robins are often on eggs in March and sometimes even earlier, when the air temperature is low enough to potentially chill uncovered eggs very quickly.

Because incubation does not begin until all the eggs have been laid, the chicks all hatch at about the same time. They use an egg tooth – a hard, button-like tip to the otherwise soft bill – to break out of the shell, but it is an exhausting and drawn-out process that not all chicks successfully complete.

Above: For the first few days the naked, blind chicks are completely helpless and need near-constant care.

Below: The chick that begs most ardently is the one that will get the food – it may be the hungriest, or just the strongest.

Life in the nest

BREEDING

Opposite: The flight feathers take longer to grow than the body plumage, and chicks will leave the nest before they are capable of strong flight.

Like most baby songbirds, newly hatched Robin chicks are feeble, blind and mostly bald. For their first few days of life their mother broods them, just as she incubated the eggs, while the male does the bulk of food collection. He may feed the chicks himself or pass food to his mate so that she can feed them.

The youngsters develop rapidly, and in their second week of life they are big and feathered enough to be left unbrooded while both parents collect food for them. Over this second week their feathers grow quickly, each appearing first as a 'pin' quill, which then breaks to reveal the soft tip of the new feather.

Below: A Robin seen carrying food is a sure sign that there is a nest with chicks nearby.

The fare brought to chicks in the nest is composed primarily of small, soft-bodied insects and insect larvae, and other squishy invertebrates. Non-spiny caterpillars and small worms are particularly important, and between them the parents bring back something in the region of 1,000 items of prey a day (sometimes carrying two or more at a time) when the young are well grown. The chicks respond to the arrival of a parent at the nest by stretching up their necks, gaping their mouths and begging with a high-pitched call. The parent feeds whichever chick is closest or begging the most insistently. It then waits a moment to see if any of the chicks are going to defecate, which they do by turning around and pointing their backsides skywards. The faeces come in a gelatinous sac, making it easy for the parent to pick up the package of poo and carry it away, to be dropped some distance away so as not to draw predators' attention to the active nest.

Fledging and independence

Robin chicks leave the nest at about 14 days old. By this time they are fully feathered, in their first or juvenile plumage, although they still have short tails and short flight feathers, and will not fly strongly for several more days. The plumage is uniformly brown with yellow tips to each feather, giving a mottled appearance. The lack of a red breast keeps the young birds safe from attack by other Robins, for the time being at least. For the next two or three weeks the chicks continue to rely on their parents for food. Often the female Robin begins a new nest before her first brood is fully independent, and the male cares for the fledglings alone for the last week or so. Fledglings also beg from any other adult Robins they encounter, although they usually get short shrift. In fact, they will beg from the same bundle of red breast feathers that has been shown to elicit attacks from adult Robins (see page 48), so right from the beginning of a Robin's life the red breast is an important signal.

Below: A newly fledged chick still has just a stub of a tail, and often a halo of baby down.

Left: The yellow 'flanges' on the mouth-sides indicate a very young fledgling, still dependent on its parents.

Below: Young Dunnocks are quite similar-looking to young Robins, but are greyer with a subtly different 'look' to the face due to smaller eyes and a different bill shape.

At first the chicks follow a foraging parent, or wait in a sheltered spot and call for food, but they quickly take an interest in what the adults are doing and begin to learn to find food themselves. As they learn, parental attention dwindles. Soon after independence the young Robins moult their juvenile body plumage. The replacement plumage is adult-like, the red breast appearing patchily and gradually over the course of a couple of weeks, although the birds retain their juvenile flight and long tail feathers for the first year of life.

Migration

We do not think of Robins as migrants, because we see them all year round. However, not only are some of our breeding Robins summer visitors only, but during winter we also play host to visiting Robins from further north and east. Migrating is a risky strategy, but for Robins, which rely so much on having a territory of their own in winter as well as summer, it can be the lesser of two evils if territories are at a premium or if severe winter weather makes day-to-day survival a huge challenge.

Where do 'our' Robins go?

In many cases adult British-breeding Robins go precisely nowhere in winter, but remain on the territory in which they bred in summer. However, a territory that can support a pair of Robins and all their offspring through summer may only be large enough for a single Robin in winter. So while in winter a breeding pair of birds may split the breeding territory between them, it is more usual for the male to keep the whole patch and drive out the female. He has little incentive to allow her to stay – if he holds his territory through the winter he is very likely to be able to attract a mate the following spring, be it the same female as the previous year or a new bird. For her part, the female may try to keep the territory for herself, but in conflicts between males and females, males tend to win by virtue of being (slightly) bigger and perhaps also because they have a stronger tie to the territory.

Ousted females may be able to establish a territory close to where they bred, or they may move away. Leaving familiar ground is

Opposite: Some of the Robins that breed here, especially females, spend the winter hundreds of miles further south.

Below: A Robin in Spain in the winter. This bird may be Spanish born and bred, or could be a migrant from the British Isles.

challenging in itself, as it generally brings them into conflict with a whole new set of Robins, so it is perhaps not surprising that many females move a long way from home, heading south to France, Iberia and even North Africa. They still meet other Robins in these countries, but they also reap the benefits of higher average temperatures and accordingly a richer supply of invertebrates to eat. However, some of our female Robins do not move very far at all, and a few do manage to hang on to their breeding territories through the subsequent winter, occasionally even ousting their male partner. There is also evidence that a few of the lighter weight males migrate, as do a fair number of first-winter birds of both sexes.

Below: We don't think of Robins as great fliers, but they are capable of covering long distances when necessary.

Where do winter Robins come from?

Because of the UK's position in the Atlantic, catching the Gulf Stream, and the proximity of everywhere in the country to the sea, it has a milder climate than areas on the same latitude in mainland Europe and Asia further east. Birds of many species take advantage of this and move across to the British Isles for the winter, while others arrive from more northerly areas. Among them are many species that we do not really think of as winter visitors – for example, a few million extra Blackbirds join our breeding population each winter.

Below: Seen in autumn, this Robin on a south-eastern English beach is probably a newly arrived migrant from the continent.

The Robins that come to Britain in winter travel from adjacent Belgium, Holland and Germany, and also from Scandinavia, the Baltic states and even as far as western Russia. Not all of the breeding Robins from these places end up in Britain, but the direction of migration is generally south-eastwards. It is difficult to assess just how many Robins arrive, because breeding birds also depart from Britain. However, on occasion there is a mass arrival of migrant birds on the east coast in autumn, and this can give a clue to the numbers involved.

Other factors that may influence the number of migrant Robins reaching Britain include how well the breeding season went (after

Mass bird 'falls'

Dramatic overnight arrivals of migrating birds are most likely to occur at east coast headlands, in poor weather conditions that compel migrants to stop and rest at the first land they find. On 22 October 2012 one such 'fall' took place on the bleak and isolated shingle spit of Blakeney Point in North Norfolk, in heavy fog, and included 280 Robins, as well as 3,000 each of Blackbirds and Song Thrushes. A similar fall at Blakeney in September 1933 was reported to have included more than 3,000 Robins.

Below: A generous crop of autumn berries provides sustenance for arriving migrant Robins and many other birds.

a good breeding season more birds compete for territory than in a poor season). A poor berry crop can force birds that depend on this resource to roam more widely than usual, and very hard winters on the continent also result in more birds being pushed our way.

How can you tell if the Robin in your garden in winter is a resident British breeder or a foreigner on its winter holiday? The short answer is that you probably cannot, unless you know for certain that you are seeing the same Robin year round because it has a unique aberrant marking or a distinctive behavioural quirk. The longer answer is that there are subtle differences – visiting migrants tend to measure a little more in wing length and weight than resident birds, and northern and eastern Robins tend to be slightly paler-toned. Behavioural differences might also be noticeable. Eastern European Robins are unlikely to be as confiding as our resident birds. However, given the species' innate curiosity and capacity to learn, even the shyest Robin from the remotest Polish woodland could be taking mealworms from a gardener's hand by January.

Below: In western Europe, Robins are largely sedentary (green shading), but further north-east they are summer visitors only (yellow shading), and move many miles south in winter (blue shading).

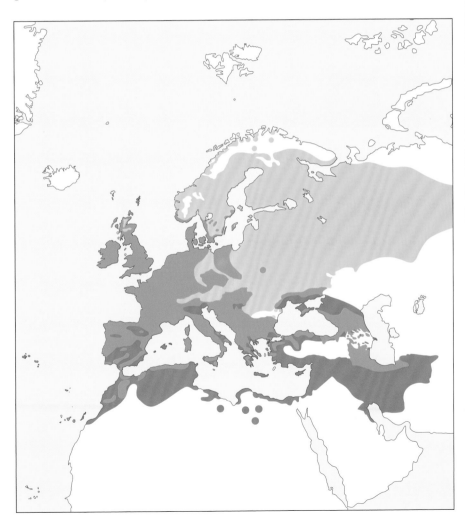

Ringing

Above: The standard BTO ring is a very lightweight aluminium band, bearing a unique number.

The study of bird migration has progressed a great deal since it was first noticed that some species seemed to vanish at certain times of the year. Early theories to explain this included the idea that Swallows spent the winter buried in mud at the bottoms of ponds and, as mentioned earlier (see page 22), Aristotle's idea that Redstarts turned into Robins in the winter. These explanations were perhaps not as daft as they seem to us now – after all, Swallows do plunge into pond water to drink and bathe, and some birds moult into distinct non-breeding or winter plumage. However, as members of our own species became world travellers, and nations communicated more and more with each other, it became apparent that a great many birds undertake long journeys twice a year, every year, to exploit favourable conditions for breeding and for surviving the winter.

The main drawback of ringing as a way of tracking birds is the numbers involved – the vast majority of ringed birds will never be found, seen or caught again. However, huge numbers of birds are ringed every year, and although the return rate is very low it has yielded enough results for us to gain insight into aspects of the Robin's

How ringing is done

By providing a means to trace the origins of particular individual birds, the advent of bird ringing opened up a new world of insights into not just the fact of migration, but its details – where the birds go and by what route, how long they take to get there, and when they begin and end their journeys. Several ornithologists tried out ways of uniquely marking birds – with a length of coloured thread tied to a leg, or a blob of ink on the feathers – but a lightweight aluminium ring around one leg soon became the standard. Such rings can be fitted securely to an adult bird's leg, can bear a unique numbered code and are durable enough to last the bird's lifetime. The science of catching birds to be ringed has also

evolved, with various complex and cumbersome traps now almost entirely replaced by the mist-net, a large stretch of fine mesh that is almost invisible when fixed across a clearing, and harmlessly holds birds until the ringer comes to pick them out.

Some ringers also ring chicks still in the nest. The same ring type can be used, and if timed correctly and done with care the disturbance to the family is minimal. The obvious advantage of ringing nestlings is that their exact age can be recorded. Ringers are skilled at picking out first-winter birds, which in the case of Robins and many other species look just like adults to the average birder. However, once a Robin is fully adult even a ringer cannot tell whether it is two, four or six years old.

Below: When birds are trapped for ringing, their 'vital statistics' (weight and a set of body measurements) are also taken.

migratory life. From ringing studies we know that some British-breeding Robins migrate to southern Spain for winter, and that a few go as far as Morocco. At the same time Robins from Poland, Lithuania, Estonia and Russia are making their way to Britain.

Some exceptional ringing recoveries include an adult (and probably female) Robin that was ringed in Kent in October 1960 and just 15 days later turned up on a ship in the Bay of Biscay, more than 803km (500 miles) away, and another ringed in Humberside that decided to spend its winter in Italy, 1,411km (875 miles) away. Birds ringed in Britain and recovered in Morocco had travelled more than 2,000km (1,243 miles). Perhaps the strangest of all is the young Robin ringed in autumn 1992 that then flew 1,231km (765 miles) north-west to reach the isolated and extremely bleak island of Jan Mayen, 600km (372 miles) north-east of Iceland.

Below: As Robins can be so confiding, it is sometimes possible to read the code of a ring without needing to recapture the bird.

Satellite tracking

The science of animal migration has taken another great stride with the advent of satellite-tracking technology. Trackers that can be fitted directly to animals are now widely used by researchers, the machinery ranging from simple devices that record changes in day length and have to be removed to recover the data, to more elaborate (and expensive) machines that transmit live positional data. This method of study presents the opportunity to map the migratory pathways of individual birds in minute detail. Bringing the weight of the trackers, especially real-time transmitters, down to the point where small and light birds like Robins can carry them with no ill effects is a challenge, and most studies to date have involved large species. However, as the technology evolves it is very likely that tracking studies on small songbirds, including Robins, will become feasible in the near future.

Migratory behaviour

We are naturally particularly curious about the mechanisms that trigger and guide bird migration. How do birds know when to set off and which way to go? Robins and many other birds carry iron-based magnetite crystals in the skin of their bills (see also page 37), which are sensitive to magnetic fields. In other bird species the magnetite has been shown to have a key role in guiding the path of migration. Experience also plays a role – migrants that have already made the journey before will use landmarks for navigation, particularly when close to their destination.

The main trigger for migratory behaviour is the change in day length. As the days shorten, a few weeks after the annual moult is completed, pre-migratory flocking behaviour can be observed in birds such as Swallows and martins. Studies on captive birds have also shown that migratory species adopt a new preferred flight direction at migration time, corresponding to the

Below: Several environmental factors, most notably the shortening days, trigger Robins to begin their autumn migration.

Above: Very severe weather can force resident Robins to make local, short-distance movements in search of better conditions.

direction they would head if free to migrate. However, a study on preferred flight direction in wild Robins in Poland, caught and released during the autumn migration period, showed less consistency compared with another species, the Chiffchaff. Unlike Robins, Chiffchaffs in eastern Europe are wholly migratory. The study's results suggest that the decisions about when to migrate and which way to go are more flexible for Robins, allowing them to 'choose' not to migrate if conditions at home are favourable.

Most small birds migrate at night, and Robins are no exception. It is sometimes suggested that they are using star arrangements as a navigational aid, but there is little evidence to support this idea. If captive Robins have their view of the sky blocked, they still orientate themselves in the correct migratory direction, but if exposed to artifical magnetic fields they change their orientation accordingly. Migrating at night is perhaps simply the safer option,

Left: Robin migration tends to be in short bursts, with frequent stops to forage.

involving less risk of being seen by birds of prey, and it leaves the daytime free for foraging and catching some sleep. Robins tend to use the 'hedge-hop' method of migration, making regular stops along the way to refuel, rather than undertaking an epic single journey. Even so, they feed up before migration and take on extra fuel in the form of fat stores, although not to the same extent as true long-distance migrants such as the similar-sized Garden Warbler, which is double its normal body weight at the start of migration.

Once on the wintering grounds our Robins in Iberia and adjacent countries do what they do best and establish a territory, which they defend from other Robins for the duration of winter. Studies suggest that the most important feature of a good winter territory for these birds is not how much food it offers, but its quality as a safe refuge from predators, which are more numerous and more varied than back home.

Life and Death

'Live fast, die young' would be an apt motto for the Robin. Its life is one of frantic activity, and its death is often quick and violent. The chances of any individual wild Robin surviving long enough to be affected by any sign of old age are vanishingly small. However, Robins have many ways of trying to cheat death and survive as long as possible. Besides feeding, breeding and defending a territory, a Robin's behaviour is directed towards keeping itself as safe and healthy as possible.

Feather fitness

A fit Robin is a bird that can take quick and effective evasive action if danger threatens, and that can survive through winter chills and lean times to take full advantage of the good feeding and warmer weather that spring brings. To fly well and keep warm as efficiently as possible, the feathers have to be kept in good order, and feather care takes up a lot of time in the average Robin's day, with preening, bathing and sunbathing all being common and important behaviours.

Opposite: When they moult, Robins are a little less beautiful and a little less resilient than usual.

The act of preening – running the feathers through the bill – does several things. The barbs along a feather shaft normally 'zip' together like Velcro, providing a wind- and water-resistant shell – preening zips up any feathers that have become unzipped. It also allows a Robin to remove any feathers that have become detached from the skin but are still caught on neighbouring feathers, and to directly remove feather parasites. Preen oil, secreted from a gland at the base of the tail, is applied to the feathers during preening and helps to keep feathers smooth, flexible and waterproof. In some bird species at least, preen oil has been shown to repel or reduce the activity of feather-damaging parasitic organisms.

Below: Careful and frequent preening helps maintain the plumage's waterproofing and insulating abilities.

Right: Robins appear to thoroughly enjoy taking a bath, but must stay alert for predators at the same time.

Below: Lots of things can play havoc with plumage, including a breezy day.

A Robin may preen in short bouts, or engage in a single long session, working its way around all its body feathers and adopting various strange postures to reach them. Feathers on the face and head cannot be preened with the bill, so instead are scratched with a foot. Robins, like many other birds, are 'indirect scratchers', lowering a wing and bringing the leg up and over it to scratch.

Bathing in shallow water is another technique employed by Robins to keep their plumage clean. A bird that is about to bathe hops to the water's edge, checking all the time for predators, and usually drinks a mouthful or two first, filling its bill, then tilting its head back to swallow. Then it moves into the water belly-deep, and bathes with vigorous dipping and flapping movements. Once the bath is completed it flies (rather heavily) to a safe, sheltered perch and preens at length.

Above: When a Robin settles down for a serious preen, it usually chooses a safe, sheltered spot.

Anting

A rare way of ridding the plumage of parasites is anting, in which a bird adopts a similar sprawled posture to that used in sunning, on or near an ants' nest. The ants, being ants, immediately climb all over the bird's feathers, squirting formic acid as they go, and this is thought to help deter feather parasites. Anting is best known here as a behaviour of Jays, but has also been observed in several other species, including Robins.

Above: Sunbathing birds will fluff up their plumage dramatically – fully separating the feathers allows the sun's warmth to reach the skin.

Sunning or sunbathing in Robins is common on sunny days. A sunning Robin looks rather stricken, crouching low with wings and tail spread and the bill gaping open. The plumage on the back is ruffled, allowing the sun's warmth to reach the skin, including the preen gland. A bird may do this simply to warm up, especially early in the morning, but sunning also helps to drive parasites from the feathers.

Moult

Feathers are marvels of engineering, with extreme strength and flexibility at a very low weight, but they do not last forever. Over time they become worn and faded, and some get broken or even pulled out before their time. Despite a bird's best efforts, some may be chewed away by feather-eating lice and mites. Adult Robins near the end of the breeding season often look particularly threadbare, with bald patches here and there, especially on their heads. Luckily, shed feathers are replaced quite quickly by brand-new ones. In late summer, after breeding, adult Robins moult and regrow all their feathers over a period of several weeks, while juvenile Robins moult their soft baby body feathers and grow their first adult-like plumage, although the juvenile flight and long tail feathers are retained for a year.

Below: Robins in fresh juvenile plumage look pristine, but very soon after fledging their first moult begins.

Above: A Robin that seems to be moulting heavily outside the season may have a case of feather mites.

Below: The post-juvenile moult begins soon after the young Robin has become independent of its parents.

The moult is a physically demanding time for a bird, but its timing (after the hard work of breeding is over, but while there is still an abundance of natural food and leafy shelter around) gives a good chance of survival. The moulting and regrowing of the long flight feathers is sequential, so the Robin is never left unable to fly, although it will not be as strong on the wing during this period as it is in full plumage. By the time that the weather starts to get colder and migratory Robins are preparing to make their journeys, the moult is over. Both adult and young Robins are now fully kitted out in a brand-new set of feathers and as ready as they can be for the rigours ahead.

Sleep

Rest and sleep are vital, but dangerous, for birds. A Robin sleeps primarily at night, when foraging is difficult and some of its most dangerous predators are also sleeping, but it is at risk of being taken by owls and cats, so must find a very secure and sheltered spot to roost. It is also vulnerable to chilling overnight – again, a well-sheltered roost helps to guard against this. A thick growth of ivy up a wall that blocks the prevailing wind is ideal. Sometimes Robins even overcome their natural antipathy towards each other to roost in groups, sharing body heat as well as increasing the chance that at least one of them will notice an approaching predator before it is too late.

Below: Robins may catnap in an alert-looking posture for a few seconds at a time. For more prolonged sleep, the head is turned and the bill tucked away.

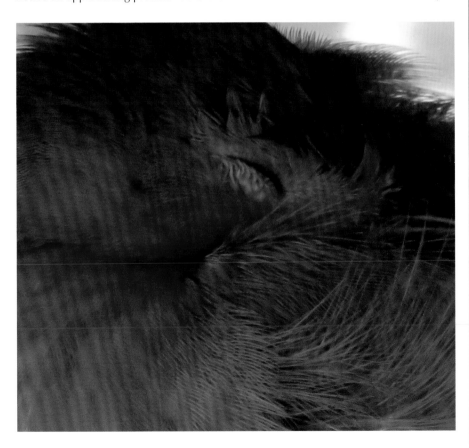

Predators and avoiding them

Robins are most at risk of being eaten when they are in the nest. Eggs and chicks alike make an easy meal for a wide range of opportunistic omnivores, from Hedgehogs, rats, Foxes and squirrels, to Magpies and Jays. Should any of these discover an active Robin nest, they are very likely to eat all the eggs or chicks it contains. Only well-grown chicks that are close to fledging stand a chance – they may be able to escape by 'exploding' from the nest and scattering into the undergrowth. Fledglings may also be taken by this category of predators, as they are still relatively weak and lacking in alertness. The same goes for adult Robins that are injured, suffering from disease or otherwise compromised.

Below: Sparrowhawks, especially males like this bird, are the most dangerous wild predators as far as Robins are concerned.

As well as their eyesight, Robins also use their ears and voices to guard against predation. Songbirds of all kinds understand each other when it comes to raising the alarm, and hearing one bird's alarm call sets off all the others and puts everyone on high alert. If a predator is spotted before it can attack, whether by its target or some other nearby bird, the resultant chorus of alarm calling results in it almost certainly failing and usually moving on. Alarm calls are generally harsh rattling, chattering or tacking notes, although many birds also have a quiet, soft whistling alarm call, used when they themselves feel in danger from a predator they have seen. Detected predators may (if no longer deemed an immediate threat) be fiercely mobbed by all the local songbirds, to chase them away from the area. This is particularly evident when a roosting owl is discovered in daylight.

Below: Avocets mobbing a Marsh Harrier. Their actions may discourage the raptor from hunting in that area in future.

Above: Small birds mainly alert each other to the presence of a predator by calls.

These types of predator pose little risk to an alert and healthy adult Robin. The range of predators that are dangerous to adult Robins is much smaller, but these are specialists at catching strong and alert prey, so are very dangerous (and even more dangerous to inexperienced young Robins). The most significant threats to adult Robins are Sparrowhawks, and (in gardens in particular) the domestic cat. In more rural settings, Tawny and Barn Owls, 'buzzing' along hedges to flush out roosting birds, present a real threat, and so do Stoats and Weasels, which are extremely fast and slim enough to chase Robins through thick cover. Robins migrating to Iberia and North Africa face additional hunters, such as Wildcats, Common Genets, various snakes and a range of additional birds of prey.

The Robin attempts to protect itself from predators by staying close to cover at all times, and always staying vigilant, pausing often while foraging to look around (and above). Both cats and Sparrowhawks use cover

Below: Sparrowhawks aren't always welcome in the garden, but their presence does indicate a healthy local ecosystem.

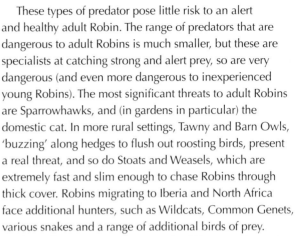

themselves to get close to their prey (or wait for the prey to get close to them), before launching a fast, short-range attack – cats hide under bushes, and Sparrowhawks fly low behind fences or hedges, or move discreetly from tree to tree. A Robin therefore needs cover that it can dive into, but that is thick enough to prevent a threat from following (or high enough, in the case of the cat). There is an ongoing struggle between the Robin's vigilance and the predator's stealthiness, and inevitably the younger and 'greener' the Robin, the more likely it is to be caught out.

Predation and population

A large-scale study on cat predation, conducted by The Mammal Society, looked at prey brought home by nearly 1,000 cats across the whole of Great Britain, during the spring and summer of 1997. Out of 14,370 items of prey, 2,809 were birds, and of those, 142 were Robins. If this result can be taken to be typical, then the Robin is the fifth likeliest bird species to fall prey to domestic cats, behind the House Sparrow, Blue Tit, Blackbird and Starling (although a large proportion, 503, of the dead birds were not identifiable). Based on the nine million-strong total British cat population at the time, the study estimated that cats kill between 25 million and 29 million birds a year, and applying this formula to the Robin tally would suggest that cats kill getting on for 1.5 million Robins every year.

Below: Domestic cats pose a significant threat to garden Robins, especially because Robins often forage on the ground.

Above: This unlucky Robin has fallen prey to a Great Grey Shrike, a predatory songbird that is a rare winter visitor to Britain.

Robins are also important prey for Sparrowhawks, forming more than 5 per cent of the hawk's diet in some studies. Sparrowhawks are specialist bird catchers, unlike cats which are more skilled at taking small mammals. It is estimated that a pair of Sparrowhawks accounts for a little over 2,000 Robin-sized birds a year, and there are about 35,000 pairs of Sparrowhawks in Britain. Even allowing for the fact that a lot of Sparrowhawk prey comprises birds much larger than Robins (up to the size of Woodpigeons, in fact), which would bring down the total number of birds taken a year, it is likely that Sparrowhawks collectively kill more than a million Robins every year.

Cat and Sparrowhawk kills of Robins therefore add up to what sounds like a great deal, but with nearly seven million Robin pairs producing up to 15 chicks each year, this rate of predation is more than sustainable, and indeed Robin numbers have been increasing at a modest rate since the 1970s.

Hunger and cold

Severe winters can take a very significant toll on bird numbers, especially when a prolonged freeze and days of snow cover make it difficult for birds to find food, and to maintain their body temperature overnight. Our small insectivorous birds which do not migrate – species like the Wren, Goldcrest and Dartford Warbler – are at the highest risk and can suffer huge population crashes through such winters. Robins, being partly vegetarian in winter, are less vulnerable, but mortality can still be high when the temperature falls very low.

Undoubtedly this is where garden feeding saves lives, and garden Robins are more likely to get through times like these than their woodland cousins. The notoriously severe winter of 1963 took a heavy toll on many bird species, including Robins, with one Devon village's estimated population of 150 pairs reduced to just 50. Of course, the survivors of such events are in a favourable position once winter is over, because there are fewer rivals competing for the available space and resources, and (providing severe weather does not continue well into spring) a bumper breeding season often follows a harsh winter.

Below: The iconic 'Robin in the snow' image belies the very real danger that such weather poses to Robins' survival.

Other causes of death

Above: Where many different birds feed together, there is a risk of infectious diseases spreading among them.

Robins are much less susceptible to outbreaks of contagious disease than other, similar-sized but more gregarious garden birds – due to their antisocial ways illness is unlikely to spread between them. However, infectious avian diseases such as chlamydiosis and trichomonosis can be spread to Robins by other birds when they come into close contact at garden feeding stations. It is important for anyone who feeds their garden birds to regularly disinfect feeders, and to cease feeding for a while if they notice any obviously sick birds at the feeding station.

Death by misadventure in the environment shared with humans accounts for a small proportion of Robins. They may drown in steep-sided ponds and troughs, be hit by cars or fly into windows. Even badly designed birdfeeders may cause injury or death by trapping a Robin by its foot or bill. Nests can be accidentally destroyed when gardeners cut down hedges – this is one garden job that should always be scheduled for outside the breeding season. Traps set for rodents sometimes inadvertently catch Robins and other ground-feeding birds, and the Robin penchant for entering buildings can lead to the birds being trapped inside with a risk of starvation if they are not noticed and released soon enough. Robins that suffer permanent injury to a wing or leg may actually

Below: Colliding with windows injures and kills many garden birds.

Above: Birds like Robins that habitually make short, low-level flights across clearings are vulnerable to being hit by vehicles.

survive quite well with a disability, particularly if they are garden Robins with a guaranteed supply of food, but when it comes to challenges like defending their territory or forming a pair bond, they are at a distinct disadvantage.

Deliberate killing of Robins by people is a very rare event. The birds and their nests are fully protected from destruction at all times whilst active. If there is ever a genuine need to kill a Robin or destroy its nest because public health and safety are at risk, a special licence to do so must be applied for and issued by the relevant government body (for example, in England special licences are issued by Natural England). Even if this protection did not exist, our national fondness for the species is very strong, old superstition maintains that harming a Robin or its nest is unlucky, and even those with no particular liking for the bird or respect for superstition are unlikely to find it a nuisance. However, in some parts of the Robin's range, particularly around the

Mediterranean, it is deliberately targeted by bird trappers and used for food – the dish *ambelopoulia*, consisting of whole cooked songbirds of various species, is a delicacy in some cultures. All EU member states are party to the Birds Directive, which gives Robins and most other bird species full legal protection, but illegal killing remains a significant problem in certain countries.

With so many dangers facing it, the Robin's average survival rate is very low. Most clutches are of five eggs, but the average number of chicks to fledge per nest is less than three. Only 41 per cent of those that manage to fledge will still be alive a year later, and for those that reach breeding age the average lifespan is a measly two years. The British longevity record for a wild ringed Robin is eight years and five months, although there are records of a 19-year-old and a 17-year-old wild ringed Robin from the Czech Republic and Poland respectively. So this little bird does have the potential to live to a ripe old age – but only if a great deal of luck is on its side.

Above: Trapping songbirds for food is an illegal but still widespread practice in Mediterranean Europe.

The Future

The continued survival of any individual Robin is on a knife-edge, with multiple hazards to be dodged every day of its life. However, as a species the Robin looks in good shape, both in Britain and across its whole global range, with most populations stable or increasing. It is easy to be complacent about such a common and adaptable bird. However, it is not so long ago that we felt the same way about the House Sparrow, yet that familiar bird has undergone a huge population crash of about 64 per cent in the UK since 1970, and we still do not know exactly why. Is there any chance that a similar catastrophe could befall the Robin?

Forests through time

Long before there were gardens to forage in and spade handles to perch upon, Robins were birds of woodland, and they still are, in rural Britain and elsewhere in their range. Pre-human Britain is thought to have been almost entirely wooded, although natural shifts in climate would have caused fluctuations in both the extent of the 'wildwood' and the types of tree that grew in it. Robins can thrive in all kinds of natural woodland, and there is every reason to suppose that they were as widespread and successful then as they are now. The earliest fossil evidence of Robins in Britain dates back to the Devensian glaciation, which took place between 10,000 and 120,000 years ago.

Human activity diminished the extent of tree cover gradually over several thousand years, a process that picked up speed halfway through the 17th century. The rapid growth of the human population, and its need for wood (for fuel and

Opposite: With its willingness to adapt from open countryside to urban habitats, the Robin is better placed to do well in the future than many other British birds.

Below: Selective tree-cutting in woodland can create a more Robin-friendly environment, with open areas for foraging.

Above: Although Robins can thrive in gardens, deciduous woodland remains an important habitat for them.

Opposite: Pine plantations are less hospitable for Robins and many other small birds, offering little low cover for nesting.

building) and cleared land (for growing crops) shrank Britain's forest cover at a dramatic rate, down to just 5.2 per cent by 1905. However, the Robin's willingness to leave the forest and live alongside people in villages and towns had been apparent long before then.

Throughout the 20th century forest cover has increased again, although initially most new woodland consisted of non-native spruce in dense plantations. The heavily shaded avenues in such plantations formed a much less hospitable environment for wildlife than native broadleaved woodland, and one in which even the ever-versatile Robin would have struggled to thrive. In 1988 the new Woodland Grant Scheme was introduced, and offered large incentives to foresters for planting native broadleaved woodland. Today, forest cover in Britain is up to nearly 13 per cent, with an increasing proportion of it being wildlife friendly. Large tracts of this 'new' woodland are still very young and have yet to develop the full species diversity of fully mature woods, but they are quite able to support good numbers of Robins, and will continue to do so as they mature.

Woodland substitutes

Right: Modern farming practices have harmed populations of many British birds.

For many woodland birds only true, dense and extensive woodland will do. Others, however, including the Robin, are able to make use of woodland edges, small copses among farmland, hedgerows, narrow belts of trees along rivers, parkland and gardens – as long as there are some trees and shrubs, they can prosper. As the UK's population grew, so the proportion of land given over to gardens and urban green spaces increased. Today about 6.8 per

Below: Leaving a strip of 'set aside' around the margins of arable fields helps support farmland wildlife.

cent of the UK is 'urban', but just over half of this area is composed of gardens, parkland and other non-concreted open space. A significant proportion of that space is potential habitat for Robins.

The value of a garden as Robin habitat depends on many factors. Some gardens offer shelter with lots of shrubs and trees, but the plants are not native so do not support many insects for Robins to feed on. Others have insect-rich patches of 'wild' meadow but no suitable nesting sites. Many people who have gardens like to feed the birds and attract them in large numbers, but non-natural food is of limited value to breeding birds. There are many resources available for people who want to make their garden as wildlife friendly as possible, but for every carefully planned wildlife garden there are five or 10 or 20 'low-maintenance' gardens with extensive concrete, decking, meticulously manicured lawns, all other plants confined to pots, and next to no resources for Robins or wildlife generally.

Above: Non-native conifers support fewer insects than native trees, making them poor foraging grounds for Robins and other insect-eating birds.

Whether Robins continue to prosper as garden birds in the British Isles depends very much on shifting fashions, and attitudes to gardens and how they should be managed. Feeding the birds is a good start, but if we want them to stick around throughout the year and successfully nest in our gardens, we need to make sure that the gardens also offer safe nest sites and plenty of habitat for insects and other invertebrates.

Farmland is not a key habitat for Robins, but where farmed fields meet woodland edge, or hedge or even a small isolated stand of trees, there may be nesting Robins. Farmland management and wildlife conservation are often uneasy bedfellows, and many farmland birds have undergone dramatic declines as farming practices have become more intense. Removal of hedgerows and use of pesticides are two activities that could harm rural Robin populations.

Climate change

Above: Cold, snowy winters are disastrous for small birds. Those that, like the Robin, exploit food offered in gardens are more likely to survive.

Our planet has undergone some very dramatic changes in climate over its long history, with corresponding changes in animal and plant populations. However, the changes of the last couple of hundred years, the consequences of global industrialisation, are cause for great concern. They are happening so fast that wildlife has no chance to adapt, and they jeopardise our own survival, too.

In a British context we are seeing the effects of climate change as more southerly species spread northwards, while more northerly species are starting to run out of road and decline. The Robin's distribution has not changed, but its annual cycle has – since 1968 the average date of the first egg laying has gone back eight days. Being able to breed earlier in the year is probably an advantage, as it extends the breeding season and means that the birds are likely to be able to have a third or even a fourth brood. Milder winters, too, are probably advantageous to Robins, as invertebrate prey is easier to find and keeping warm overnight is less of a struggle.

Left: Heavy rain in early spring can delay breeding as Robins struggle to find enough insect prey.

In the longer term, certain predicted consequences of climate change could be detrimental. Many climate-change models predict that more extreme weather patterns are likely, bringing a higher risk of both drought and heavier-than-usual rainfall. Both of these can have serious impacts on breeding success.

Below: A fledgling Robin in north Scotland. The most northerly breeding Robins produce fewer chicks, because of the shorter summer season.

Alien invasion

One of the most severe threats to wildlife on a global scale is that posed by non-native species, introduced (deliberately or accidentally) to a new country by people. In some cases they integrate with native wildlife and cause few problems; in others they are unable to survive well and quickly die out. However, if they survive well they often do so to the detriment of one or more native species, which they prey on, outcompete for resources or infect with diseases to which the native species has no immunity. Wildlife communities that evolved on isolated islands are most at risk, as often they are totally without effective defences. A namesake of the Robin, the Chatham Island Robin, was brought to the brink within decades by introduced cats and rats on its native Chatham Islands, off the east coast of New Zealand; the predators found the confiding and weak-flying little bird exceptionally easy prey, and only very robust conservation efforts saved it from extinction.

In Britain there are quite a number of successful introduced species, although few have an obvious direct

Below: Non-native predators almost wiped out the Chatham Island Robin – only good luck and great conservation effort saved it.

impact on Robins. The Little Owl, introduced in the mid-19th century, probably preys on Robins occasionally, but the two species coexist on mainland Europe, and in any case the owl is primarily an insectivore. The American Mink, descended from fur-farm escapees, can have a serious impact on bird breeding success, but it is mainly found in wetland habitats so is not a serious danger to Robins. The Grey Squirrel does share habitat with Robins and is an opportunistic nest predator, but so is the Red Squirrel that it has replaced across much of Britain, so the net change for Robins is probably minor. None of the non-native birds that have become established in Britain are sufficiently Robin-like that they could pose a serious danger to our Robins in terms of competition for food, nesting sites and other resources.

Today there are strict controls in place preventing introductions of alien species, but it is all too easy to imagine horror scenarios – for example, the Brown Tree Snake of Australia has had a devastating impact on small-bird populations on the island of Guam, in Micronesia, where it was introduced in the mid-20th century. Prevention is much better than cure when it comes to managing invasive species.

Above: Of all the established non-native birds in Britain, only the Little Owl is a potential predator of Robins.

Below: Brown Tree Snakes have caused the extinction of 12 bird species on the island of Guam, to which they were accidentally introduced in the mid 20th century.

The Robin in its ecosystem

While Robin numbers are holding fairly steady in Britain, this is not the case for most other species. In 2013 the RSPB and other nature conservation bodies published a wide-ranging report, *State of Nature 2013*. This report presented population trend data for 3,148 species of animal and plant, from birds to bees, fish to flowers and moths to mosses, gathered over several decades. Of these species, 60 per cent have declined over the study period – and 31 per cent have declined severely.

Below: Despite its small size, the Robin is close to the top of the food chain as a predator of many kinds of invertebrate animals.

Most of the declines are primarily associated with habitat loss. Ours is a crowded island, and some particular habitat types have suffered disproportionately

high losses – for example, we have lost 97 per cent of our lowland meadows since the 1930s, 44 per cent of Scottish peat bog since the 1940s and about 80 per cent of lowland heath over the last 200 years. Moreover, while woodland cover has increased overall in the last 100 years, certain particular subtypes of woodland have not – for example, some 90 per cent of coppiced lowland wood has gone over the same time period.

As a highly generalist species, the Robin is insulated to some degree from these changes. However, the declines are so wide-ranging and so severe that if things do not change even Robins will be detrimentally influenced by them.

Below: Reflecting on the future – Robins are in good shape in Britain at the moment, but as more and more green spaces are lost, their options become increasingly restricted.

Above: Gardens with no lawn or soil, and plants confined to pots, offer precious little foraging ground for birds like Robins.

Robins in Culture

The British Isles is home to nearly 300 regularly breeding bird species, but many of them are never seen by the average person, and even among the common species, most go unnoticed for most of the time, particularly smaller birds. The Robin, however, is not easy to ignore. We encounter it constantly – in the flesh in our parks, gardens and woodland, and in writing as a character in some of our most beloved tales and legends; its image is everywhere we look at Christmas, which seems to begin earlier every year. No wonder it is one of the best-known British birds – perhaps the best known of all.

Early mentions

Before the 15th century the Robin was known as the 'redbreast' in England, and sometimes as the 'ruddock', both names referencing its most obvious feature (and it is worth noting that the word 'orange' did not exist at that time). In his poem *Parlement of Foules*, written sometime in the 14th century and telling the tale of love-rival eagles, Geoffery Chaucer referred to it as the 'rodok' (and made mention of its tameness).

'Redbreast' was extended to 'Robin Redbreast' due to a 15th-century habit of adding a 'people' name to familiar birds' names – other examples include 'Philip Sparrow', 'Tom Tit' and 'Jenny Wren'. The rare use of a female nickname for a bird was because – in stories if not in reality – Jenny Wren was seen as Robin Redbreast's mate. However, only the Robin is now known universally by its human nickname alone, the BTO having changed its name from Robin Redbreast to Robin on the official list of British birds in 1952.

Opposite: An ice-skating (and oddly human-legged) Robin features on an 1870 greetings card.

Below: The 15th century habit of giving familiar bird species human nicknames has left its mark in modern culture.

Above: Robins turn over woodland leaves to look for food, and (in stories at least) to provide lost children with a cosy bed.

One of the earliest well-known references to Robins in literature is in the 16th-century story *Babes in the Wood*. Much revised and retold over the centuries, this story in all its forms features a Robin in its final, rather unhappy scenes. The two eponymous children suffer a series of disasters, beginning with the deaths of their parents and ending when they become lost in the woods after being kidnapped by 'ruffians' hired by their evil uncle. They wander the woods for a while but eventually lie down and die, and a thoughtful Robin comes to lay leaves and flowers over their bodies. The story is perhaps inspired by the Robin's habit of turning over leaves while searching for food on the woodland floor.

This notion of the Robin as kind and compassionate is reflected in another well-known story which, although it concerns events from the Bible, does not actually appear in biblical verses. The legend tells that, moved by Jesus's plight as he went to be crucified, a small bird flew down and tried to pull away the crown of thorns on his head. In the process she became bloodstained, either with Jesus's blood or her own, from a thorn puncture, and so the red-breasted Robin came into existence.

The works of Shakespeare are well known for avian name dropping, and the Robin makes several appearances. In *Two Gentlemen of Verona* the servant

IN ONE ANOTHER'S ARMS THEY DYED.

Speed tells his master Valentine that he knows Valentine is in love because 'you have learned (like Sir Proteus) to wreathe your arms like a malcontent; to relish a love-song, like a robin-redbreast'. And in *Cymbeline* the events of *Babes in the Wood* are repeated – a speech by Arviragus makes mention of a Robin tenderly laying flowers upon the corpse of Fidele, the man he loved (who in fact turns out not to be dead, and not to be a man).

Above: Only in fiction could so many Robins gather in one spot without territorial violence breaking out!

The Robin in poetry

'A robin redbreast in a cage puts all Heaven in a rage' is a well-known line of verse from *Auguries of Innocence*, written in 1803 by William Blake. The poem's intent is to juxtapose innocence and evil in a series of paradoxes – it seems that to Blake, the idea of caging a wild Robin represented the curtailing of all freedom and was almost unthinkably evil. William Wordsworth also had a fondness for Robins and his works include a couple of poems all about them. In one, *The Redbreast Chasing the Butterfly*, he talks affectionately of the 'pious bird with the scarlet breast', the bird 'whom man loves best', but he goes on to chide the Robin for pursuing a butterfly, arguing that the two should be friends.

The nursery rhyme 'Who Killed Cock Robin' has been around in one form or another since the mid-18th century. The killer is revealed to be the sparrow, with his bow and arrow, and the rhyme goes on to describe which bird or other animal will take responsibility for the various aspects of Cock Robin's funeral and burial, with the murderous sparrow being hanged for his crime in some versions. Various theories link the rhyme to certain actual historical events, such as the fall from power of 18th-century Prime Minister Robert Walpole. Another, much shorter nursery rhyme, which dates from the 16th century, is 'The North Wind doth Blow', in which a Robin ('poor thing') waits out the cold and snowy winter by hiding in a barn.

Below: 'The North Wind doth Blow' is a charming 16th-century nursery rhyme, which recognises that however hard winter may be for us humans, it's much harder on our favourite garden bird.

Opposite: *The Death of Cock Robin*, this painting shows a scene from the well-known nursery rhyme.

The Robin at Christmas

Opposite: A Christmas card from the late 19th century casts a Robin in the role of postman.

Robins amid snowy scenery feature very heavily on the Christmas cards we send each other, and have done since Victorian times, when red-uniformed postmen were nicknamed 'robins'. Today the range of Christmas-related merchandise we can buy has expanded dramatically, but Robins still take centre stage, appearing on advent calendars, as decorations for Christmas trees and Christmas cakes, and as part of the pattern on knitted Christmas stockings.

Robins may even pop up in Christmas card images of the Nativity, as legend tells that a Robin entered the stable where the baby Jesus lay in his manger, and beat its wings to keep the manger-side fire going. In the

process, its breast plumage caught light, and thus yet another explanation for the red breast came into being (although once again the Bible does not mention it). However, in Britain Robins were associated with winter festivals long before the advent of Christianity in the country. The Winter Solstice celebrates the dawn of a new Waxing Year and the Robin Redbreast is its king, taking over from (by killing) the Wren, king of the outgoing Waning Year.

Above and right: Special Christmas stamps featuring Robins are regularly issued by Royal Mail.

A Merry Christmas and a Happy New Year.

The Robin in modern times

Above: The RSPB has long used Robin-shaped coin collection boxes at its nature reserves, as an appealing way to gain extra funds.

The RSPB's logo features an Avocet, but many of the large collection boxes that the organisation uses on reserves for donations are topped by a giant plastic Robin with a coin slot in its head. While a Robin is probably much easier to cast in plastic than a long-necked, long-billed Avocet, the universal familiarity of the Robin is another good reason to use its image in this way.

Best-known British bird?

In 2005 researchers at Newcastle University showed images of a range of British birds to 200 schoolchildren, and asked the children to name the birds. The results were widely quoted in newspapers as evidence of the modern-day disconnect between young people and nature, with most of the children unable to put a name to the likes of Blackbirds, Kestrels and Starlings.

However, every single child recognised the Robin. All of those Christmas cards must partly explain this, but due to the Robin's abundance and approachability, surely most of these children would have seen Robins in life as well as on cards – something that could not be said for another bird that the children also recognised – the Puffin.

Our universal fondness for Robins came to the fore in 2014, when Natural England announced a proposal to add the Robin to the general licence. The proposal made it legal for landowners to destroy Robin nests without applying for a special licence or any record being made of the destruction, if the nest's location threatened public

Above: Not all garden wildlife is welcomed by the garden's owners, but Robins are almost universally loved.

health and safety. The public outcry in response to this was tremendous, with an online petition against the proposal quickly amassing more than 100,000 signatures. While Robins do sometimes nest in highly inconvenient places, situations where a nest poses a genuine danger to public health and safety must be very few. The RSPB formally objected to the proposal, arguing that where there is a genuine concern the existing provision to apply for a special licence to remove the nest is sufficient.

Below: As the law stands it is illegal to disturb an active Robin's nest, and most householders learn to live with indoor-nesting Robins.

Science and nature

While the Robin's appearances in verse and fiction may be full of whimsy and sentiment, it has also received much attention over the years from ornithologists, with the result that it is one of Britain's best-understood bird species. Due to the pioneering work of David Lack (see page 43), most of us know that the Robin is actually a fierce and feisty bird, willing to battle its own kind to the death in disputes over territory. Lack's monograph *The Life of the Robin* remains a classic work of popular science more than 70 years after its first publication. More recently biologists have uncovered the secrets of the Robin's DNA, its hormones, how it learns to sing and how it finds its way on migration.

Below: Decades of study on Robins, through ringing work and other methods, has yielded many valuable insights into the species.

Above: For many urbanites, watching the garden Robin is a rare and much appreciated chance to connect with nature.

The Robin in Britain has opted to live closely alongside humans, and historically we have not been too keen on wildlife doing this – some of the most mistrusted, feared and generally disliked animals of all are those that live in association with people. However, in the case of the Robin, for once we have repaid this trust not by reviling, persecuting or exploiting it, but by forging a relationship based fundamentally on mutual benefit. We would do well to apply that same principle to how we treat the natural world in general, and learn to love the insects on which Robins prey, the hunters that prey on Robins, and all the other species that share its world and ours, just as much as we love the Robin itself.

Above: For many urbanites, watching the garden Robin is a rare and much appreciated chance to connect with nature.

Glossary

Australasia Australia, New Zealand, New Guinea, and the neighbouring islands of the Pacific.

brood The group of chicks that have hatched out from a clutch of eggs.

call Simple sound made by a bird used as communication with others of its species.

chats Several closely related, small, thrush-like bird species, including the Robin.

clutch Set of eggs that are laid on successive days and incubated together.

fledge To leave the nest and fly for the first time.

fledgling Young bird that is fully feathered and has left the nest.

flight feathers Long wing feathers, the outermost being primaries and the innermost secondaries.

foraging Searching for food.

genus (plural genera) Group of very closely related species.

keratin Tough but very lightweight natural protein, from which birds' feathers and bills are formed.

migration Journey undertaken in autumn (and back again in spring) between the breeding area and a wintering area.

moult Process of shedding and regrowing feathers.

nestling Young bird that is not yet fully feathered and is still in its nest.

plumage A bird's complete 'coat' of feathers.

primaries Outermost long wing feathers.

song Set of distinct, often complex vocalisations birds (mainly males) make to proclaim their territory and attract a mate.

taxonomy Study of how different kinds of animal and plant are genetically related to each other.

territory Area of habitat that a bird lives in and defends from others of its own species.

Further Reading and Resources

Dominic Couzens, *The Secret Lives of Garden Birds*, Christopher Helm, London, 2004

Mark Golley and Stephen Moss, *The Complete Garden Bird Book*, New Holland, London, 2011

David Lack, *The Life of the Robin*, Witherby, London, 1943

Chris Mead, *Robins*, Whittet Books, London, 1984

BTO (British Trust for Ornithology) www.bto.org

IUCN (International Union for Conservation of Nature and Natural Resources) www.iucn.org

The National Trust www.nationaltrust.org.uk

RSPB (Royal Society for the Protection of Birds) www.rspb.org.uk

The Wildlife Trusts www.wildlifetrusts.org

Acknowledgements

I would like to thank Simon Papps for initially commissioning this project, and Julie Bailey and Alice Ward for seeing it through to publication. Thanks to Krystyna Mayer for editing my text and making many improvements, and to Rod Teasdale for creating the pleasing page design. A number of photographers – some professional and some merely talented – have supplied images for this book, and my thanks go to them all for their contributions. Finally, I would like to thank my friends and family for their support and understanding while I was working on this book, and the many scientists and amateur ornithologists whose observations have provided so much insight into the lives of these wonderful birds.

Image credits

Bloomsbury Publishing would like to thank the following for providing photographs and for permission to reproduce copyright material.

While every effort has been made to trace and acknowledge all copyright holders, we would like to apologise for any errors or omissions and invite readers to inform us so that corrections can be made in any future editions of the book.

Key t = top; l = left; r = right; tl = top left; tcl = top centre left; tc = top centre; tcr = top centre right; tr = top right; cl = centre left; c = centre; cr = centre right; b = bottom; bl = bottom left; bcl = bottom centre left; bc = bottom centre; bcr = bottom centre right; br = bottom right

AL = Alamy; FL=FLPA; G = Getty Images; RS = RSPB Images; SS = Shutterstock

Front Cover t SS, b SS; **Back cover** t SS, b SS; **1** Mike Lane/RS; **3** Peter Cairns/RS; **4** Chris O'Reilly/RS; **5** Kristin McKee/G; **6** Ray Kennedy/RS; **7** t Marianne Taylor, b Ann & Steve Toon/G; **8** Harvey van Diek/FL; **9** Rick van der Weijde / Pbase.com/corotauria; **10** t SS, b BlackCatPhotos/G; **11** Robin Chittenden/FL; **12** t SS, b SS; **13** t Paul Dean, b Paul Dean; **14** FL; ImageBroker/Imagebroker/FLPA; **15** tl Mark Sisson/FL, tr Roger Wilmshurst/FL, cr Dave Pressland, bl Erica Olsen/FL; **16** Nigel Blake/RS; **17** Malcolm Schuyi/FL; **18** Michael Roberts/G; *19* Andrew Parkinson/RS; **20** t Marianne Taylor, b Oxford Scientific/G; **21** Marianne Taylor; **22** t Marianne Taylor, b Marianne Taylor; **23** t Marianne Taylor, tc Mike Lane/RS, bc SS, b Marianne Taylor; **24** SS; **25** SS; **26** SS; **27** t Martin Willis/FL, b SS; **28** John Cancalosi/G; **29** Mike Lane/FL; **30** Kevin Schafer/G; **31** SS; **32** SS; **33** Andrew Bailey/FL; **34** SS; **35** t Lizzie Harper, b Lizzie Harper; **36** David Hoskins/FL; **37** SS; **38** David Hosking/FL; **39** Mike Read/RS; **40** Steve Austin/RS; **42** Marianne Taylor; **43** Mike Lane/FL; **44** Ed Drewitt; **45** Marianne Taylor; **46** SS; **47** SS; **48** SS; **49** Roger Wilmshurst/FL; **50** Ray Kennedy/RS; **51** Ray Kennedy/RS; **52** Marianne Taylor; **53** SS; **54** t Jack Perks/RS, b Roger Wilmshurst/FL; **55** Marianne Taylor; **56** t SS, b Ray Kennedy/RS; **57** t Marianne Taylor, b SS; **58** Steve Round/RS; **59** Andrew Bailey/FL; **60** SS; **61** SS; **62** SS; **63** Marianne Taylor; **64** t Mike Jones/FL, b Mike Jones/FL; **65** WildPictures/AL; **66** Derek Middleton/FL; **67** t SS, b Cultura RM/Oanh; **68** Marianne Taylor; **69** Roger Wilmshurst/FL; **70** SS; **71** t SS, b Marianne Taylor; **72** Erica Olsen/FL; **73** Marianne Taylor; **74** blickwinkel/AL; **75** Marianne Taylor; **76** John Hawkins/FL; **77** Julie Dando/Fluke Art; **78** Marianne Taylor; **79** SS; **80** Marianne Taylor; **81** Imagebroker, Marko König/FL; **82** David Broadbent/RS; **83** Paul Sawer/FL; **84** Peter Mulligan/G; **85** Peter Szekely/AL; **86** t Marianne Taylor, b Marianne Taylor; **87** Marianne Taylor; **88** Marianne Taylor; **89** Marianne Taylor; **90** t SS, b Dr T J Martin/G; **91** Marianne Taylor; **92** SS; **93** t Nigel Blake/RS, b Roger Tidman/RS; **94** SS; **95** Marianne Taylor; **96** Duncan Usher/FL; **97** Marianne Taylor; **98** Erica Olsen/FL; **99** Konrad Wothe/FL; **100** Fabio Pupin/FL; **101** Robin Chittenden/FL; **102** Paul Miguel/FL; **103** SS; **104** David Broadbent/RS; **105** Nicholas and Sherry Lu Aldridge/FL; **106** t Gary K Smith, b Erica Olsen; **107** SS; **108** SS; **109** t SS, b Marianne Taylor; **110** Robin Bush/G; **111** t SS, b Michael & Patricia Fogden/G; **112** Marianne Taylor; **113** t SS, b SS; **114** Amoret Tanner/AL; **115** Tony Watson/Al; **116** Michel Geven/FL; **117** The Print Collector/AL; **118** Bob Thomas/Popperfoto; **119** UniversalImagesGroup/G; **120** t SS, b SS; **121** Amoret Tanner/AL; **122** Andy Hay/RS; **123** t Gianpiero Ferrari/FL, b David Hosking/FL; **124** SS; **125** SS;

Index